MW01070435

To Keith

Preface

Statisticians often work with people in other professions and disciplines. Working with others and exchanging ideas across disciplinary boundaries is fascinating, challenging, and also frustrating at times. To some statisticians, the ability to communicate well with their clients seems to come very easily and naturally. They can establish good, collaborative relationships, elicit the details of a client's project, explain statistics clearly, adapt to differences in style, and respond constructively to problems. The rest of us could probably benefit from some training in these communication skills. This training can start at the graduate level with a course on statistical consulting. From there, a statistician can gain further benefit with additional training while on the job. The purpose of this book is to teach communication skills to statistical consultants. The book provides guiding principles, memorable illustrations, examples drawn from the actual experiences of statistical consultants, exercises, and suggestions for instructors. The accompanying video provides dynamic illustrations of both good and poor communication skills. An instructor can use this book and video to organize a graduate-level course in statistical consulting. It is also possible to organize short courses or workshops from excerpts of these materials, targeted to statisticians who are already on the job.

I believe that statistical consulting is best learned by a combination of training, observation, practice, feedback, and discussion. I encourage instructors to minimize didactic lecture in a statistical consulting course. Instead, the goal of this book and video is to provide resources and opportunities for learners to develop their own effective consulting styles. There are exercises in each chapter that direct the student to obtain information, ideas, and feedback from experienced statisticians and clients. There is a section called *Suggestions for Group Discussion* at the end of each chapter. This section includes suggestions on how to organize opportunities and resources for a class or other discussion group. Statisticians who are working through this material on their own and would like to learn more from their clients and colleagues may also use these suggestions.

I first became interested in statistical consulting when I took a consulting course in graduate school. I continued to develop my consulting skills while on the job. I realized that I needed to improve my communication skills in order to work more effectively with people from other disciplines. When I began teaching the statistical consulting course at Penn State University, I discovered that there were very few resources on this topic that were directed to a statistical audience. I was very fortunate to be able to collaborate with Dr. Ann Greeley, a psychologist with a specialty in communication. Together we produced a videotape called *Communication Skills in Statistical Consulting*. I was also able to collaborate with Dr. Sandra Stinnett, a statistician who has worked in the field of statistical consulting for many years. Together we developed a one-day short course on communication skills, with the sponsorship of the American Statistical Association. These experiences as a statistician and teacher have provided the motivation, the experiences, and the philosophy that have shaped this book. I believe that the best way for statisticians to learn these skills is to have some formal training within their graduate programs, and then continue developing their skills through the course of their careers. For this reason, I have structured the materials so that they may be used in a graduate level course, in a short course for statisticians on the job, or as a resource for statisticians who wish to learn these skills on their own.

Coordinating the Book and the Video

Video is a good medium to show the dynamics of a consulting session. The conversation and the body language of the consultant and the client can be represented in a video much more effectively than on the printed page. When statisticians are able to see both "poor" and "good" examples of the same skill on video, they can have a good laugh at the actors and remember to pay better attention to their own skills.

The video that accompanies this book portrays a statistical consultant and a client in a series of meetings. The video consists of brief dramatized segments that illustrate critical consulting moments during the course of a hypothetical project. Segments vary in length from one to seven minutes. A narrator introduces each segment and then offers some interpretation at the end of the segment. This gives continuity to the story line. This method is used to depict the process of statistical consulting that actually would have unfolded over a much longer period of time.

Chapters 3 through 8 of the book are coordinated with the unfolding story of the video. In several of these chapters, the student is directed to view a certain segment of the video before reading further. The chapter then continues with a discussion of the video. Instructors can organize the presentation and discussion of the video in several ways, depending on the needs and resources of their courses. These recommendations appear in Suggestions for Group Discussion at the end of each chapter. This section also includes suggestions on how to augment a course with recordings from actual consulting sessions.

Assessment

There are very few absolutes when it comes to statistical consulting, and there are a diversity of consulting styles that are effective. Statisticians on the job do not take written examinations in statistical consulting. Instead, they depend on feedback from clients, peers and supervisors. As an instructor, I found that providing students with feedback from their own clients and from their peers was far more motivating than any letter grade that I could assign. I also found that when students worked together to develop standards of professional conduct, they maintained these standards consistently in their work with clients. In a course on statistical consulting I believe that an instructor should place less emphasis on letter grades and test scores and more emphasis on motivation, coaching, peer review, and client feedback. In this way the instructor will reflect the form of assessment that a student will encounter in his or her future work environment.

My Perspective on Statistical Consulting

There are three themes that form the perspective of this book. First, I believe that a statistician should treat each client as a potential collaborator. This emphasizes the need for all parties to feel satisfied with the consulting in a long-term sense. It also motivates the need to treat everyone with respect, to negotiate towards "win–win" outcomes, to give and receive feedback and to resolve difficulties to everybody's satisfaction. The second theme is an appreciation for diversity. Diversity as it affects statistical consulting takes on many forms. By definition a statistical consultant will be working with people from different disciplines. There will also be a variety of cultures, styles of communication and negotiation, preferences for learning, and other dimensions of diversity that will all have an impact on the communication between consultant and client. This is why a statistical consultant needs to know how to recognize these differences and adapt to them. The third theme concerns the involvement of experienced statisticians and clients in the process of training statistical consultants. Every chapter includes suggestions on how to involve experienced statisticians and clients in the activities of the chapter. I believe that all parties will benefit from learning more about each other's points of view. This process should also benefit the statistics profession as a whole as we develop a more informed and unified approach to the training and practice of statistical consulting.

Acknowledgments

Many people have contributed, either directly or indirectly, to the effort that went into producing this book and video. Special thanks are due to J.L. Rosenberger, S.F. Arnold, and W.L. Harkness for giving me the opportunity to become a statistical consultant and to teach statistical consulting. I am also grateful to the many clients who taught me how to communicate more effectively. There are too many to name individually, but I would especially like to mention P. R. Cavanagh, Roy Hammerstedt, and P. M. Kris-Etherton. I also learned a great deal from the graduate students who took the statistical consulting course and worked at the consulting center. In particular, Lisa Suchower, Kate Meaker, Jean Recta, Brenda Gaydos, John O'Gorman, and Sandy Balkin all gave me helpful feedback. Colleagues in statistical consulting, including Ron Wasserstein, C. Chatfield, D.A. Zahn, Brian Yandell, and Ian Gordon provided many insights based on their research and experiences.

Sandra Stinnett encouraged me to embark on this project and gave helpful feedback on early drafts. Ann Greeley provided me with invaluable advice through the years about communication skills, and I am glad that she was available to narrate the video. Gary Purdue, Frank Wilson, and their crew produced the high quality video and contributed many good ideas to its production. Charles Dumas was a wonderful improvisational actor.

I changed jobs in the midst of writing this book, and I had the opportunity to get to know a lot of new clients and statistical colleagues just while I was focusing on this book. My thanks go to

Anna Nevius and to many other review scientists at the Center for Veterinary Medicine within the US Food and Drug Administration[†] for giving me a fresh perspective.

I appreciate the support that Duxbury Press has given to this project. Thanks are due to Carolyn Crockett and all of the staff who participated in its development. I would also like to thank the reviewers who provided feedback on the developing manuscript: Thomas B. Barker, Rochester Institute of Techology; Ann P. Cross; Michael Daniels, Iowa State University; Marcia Gumpertz, North Carolina State University; William I. Notz, Ohio State University; Larry J. Ringer, Texas A&M University; Robert L. Schaefer, Miami University; Roy Tamura, Eli Lilly and Company; Mark van der Laan, University of California, Berkeley; Steven J. Verhulst, Southern Illinois University School of Medicine.

Finally, I would like to express my thanks to my parents, Donald and Helen Derr, for giving me a love of learning, and to my husband, Keith Ord, for his patient support, advice, and many good ideas.

Janice Derr

[†] The opinions and information in this book are those of the author and do not necessarily reflect the views and policies of the FDA.

Brief Contents

Brief Contents

Contents

nsulting

Communication

1

INTRODUCTION

1.1 Why Does a Statistical Consultant Need to Have Good Communication Skills?

If you are planning to do any statistical consulting during your career as a statistician, then you will be working with people in other disciplines and professions, helping them to understand how good statistical practices can improve the quality of their work. Congratulations on your choice of profession! Statistical consulting can be stimulating and rewarding. It can also at times be very frustrating. In short, statistical consulting is challenging enough for a lifetime of learning.

Exhibit 1.1 ***The core process of statistical consulting***

Communication is what links you and your statistical knowledge with your clients and their statistical needs. Good skills in communication will enable you to apply your statistical training effectively to problems arising in other fields. At the very core of statistical consulting is the translation of your client's problem into a more abstract statistical representation (Exhibit 1.1, arrow *a*). The relevance of this translation depends on how clearly and accurately you have understood the problem. This is where communication skills come in. The statistical solution you develop will be based on this abstraction (Exhibit 1.1, arrow *b*) and will reflect your training in statistics. Once you have developed a recommendation, you will then translate it back to the client (Exhibit 1.1, arrow *c*). The extent to which your client understands and accepts your recommendation also depends on your communication skills. This is why we say that skills in communication are *enabling* skills; they enable you to make the best use of your expertise in statistics.

The purpose of this book is to help you master skills in communication that promote effectiveness in statistical consulting. I believe that "effectiveness" is the result of two outcomes: (1) both you and your client are satisfied with the experience, and (2) you and your client have made use of good statistical practices. Of course, the exact definition of what good statistical practices are will vary among statisticians. The focus in this book is on the communication skills that can promote these two outcomes. It is difficult to be effective if your client does not feel comfortable talking with you, if you have not correctly understood what his project is all about, or if your client does not understand what you are telling him about statistics. In order to become an effective statistical consultant, you will need to master the core communication skills shown in Exhibit 1.1. You will also need to master other communication skills required to establish and maintain a good working relationship with your client. Being able to recognize and adapt to your client's preferred style, identify the priorities of a project, negotiate a satisfactory consulting arrangement and respond to difficult situations will all help establish this relationship. That is what this book is all about.

The words "statistical consultant" and "client" are not perfect. To some people they imply an unequal, non-collaborative relationship. However, I believe that you will be in the best position to promote good statistical practices in fields of application when you have a satisfying, collegial association with the non-statisticians on a project. In this book I will use the term "statistical consulting" to represent the entire range of involvement that you can have with these projects. I will use the term "client" to represent the person, usually a non-statistician, to whom you must communicate statistical information. In general, your client originates the problem to which you will be applying your statistical knowledge. In some situations your involvement with a client's project might be very limited. Someone might call you up on the telephone and ask you for the answer to "one quick question." In other situations your involvement might be very extensive. You may work with a client from start to finish on a series of projects. I believe that good skills in communication will help you to establish a satisfying association with your clients, regardless of your level of involvement in their projects.

Finally, if you still need some extra persuasion that statistical consultants need good skills in communication, here is an excerpt from a commentary titled "Statisticians and Communication", written by Jonas Ellenberg, the 1999 President of the American Statistical Association:

> A common thread runs through the types of negative comments we hear about [clients'] experiences with statisticians. The threads that bind are our less-than-adequate communication skills. By communication, I mean an understanding and exchange of information. In general, the statistician needs to bring or derive a reasonable understanding of the problem at hand in order to make a meaningful contribution. This is accomplished by listening and hearing, digesting and responding. Some people work this way naturally, but most do not. As statisticians, we do not receive formal training in the tactics of communication. ...
>
> Better communication will likely make us stronger collaborators. Assuredly, and perhaps more importantly, it will raise the level of respect with which statisticians are viewed. As with most activities in life, while training and competence are a necessary condition for success, they are not sufficient. Respect breeds confidence – both self-confidence of the individual statistician and confidence of the collaborator in the statistician – and confidence will open doors for higher levels of collaboration ... Training in communication skills should begin during or immediately following statistical training.[1]

So congratulations again – by making a commitment to improving your skills in communication you are not only helping yourself become a more effective statistical consultant, you are also helping to improve the public image of the statistics profession. This improved image will help to maximize the rewards and minimize the frustration for statisticians and the people they work with.

1.2 What Communication Skills Are Necessary for Effective Statistical Consulting?

In this book you will find many examples of communication in statistical consulting, some successful and others less than successful. These examples are drawn from the experiences of statisticians in many different work environments. Some of these statisticians work in industry, some in government, some in an academic setting, and others work independently. In the video that accompanies this book, I play a statistician named "Dr. Derr." An actor plays a client named "Brian Johnson[†]." Mr. Johnson is the manager of quality services at the medical center in

---

[1] From "Statistics and Communication" by J. H. Ellenberg, pp 8-9. Copyright © 1999, American Statistical Association.

[†] The names (except for "Dr. Derr") and details of the video story line are fictitious.

a university setting (known as "UMS" in the video). He wants to survey the population of student users in order to assess the level of satisfaction with UMS services. The video consists of 11 segments that are brief excerpts of discussions between Dr. Derr and Mr. Johnson during the course of this project. The purpose of the examples in the book and the storyline of the video is to illustrate the communication skills that are important in statistical consulting. These are:

Communicating non-verbally: In Chapter 3 you will learn about how nonverbal communication influences the outcome of a consultation. The video will show you two "wordless" versions of a meeting between Dr. Derr and Mr. Johnson, one of them with positive and the other with negative nonverbal communication. You will find examples that illustrate how to recognize and adapt to cultural differences in what people expect non-verbally. You will also consider strategies for conveying these non-verbal messages in the variety of media in which statistical consulting takes place.

Conducting a meeting. Statistical consultants attend a lot of meetings! In Chapter 4, you will learn about different preferred styles of communication that that people may have. These preferences affect the way that a meeting is structured. You will find examples that illustrate how to participate effectively in the meetings, including one-on-one meetings, team meetings, meetings in person, meetings by remote media, and meetings in which you are the leader. The video depicts a meeting between two people with distinctly different preferences for structuring information. You will see how Dr. Derr manages to adapt to Mr. Johnson' preferred style to get the information she needs.

Asking good questions. This is one of the core communication skills in statistical consulting. Many statisticians, even experienced ones, worry that they won't know what questions they should ask when they meet with a client for the first time. In Chapter 5, you will learn about the three main types of investigation and the statistical issues that are typically associated with the design and analysis stages of each type. This will provide a general framework for you to decide what you need to find out about a client's project. The focus of Chapter 5 then turns to how you can obtain this information accurately. You will find numerous examples that illustrate these skills in different settings. The video provides both a positive and a negative version of Dr. Derr attempting to learn about the survey that Mr. Johnson wishes to conduct. You will be able to evaluate the quality of information that Dr. Derr obtains in each version.

Negotiating a satisfactory exchange. At the heart of statistical consulting is an exchange: You agree to do something for the client and in exchange, the client will do something for you. These exchanges can become complex in statistical consulting, as you will learn in Chapter 6, because both tangible and intangible items are usually involved. You will learn how to identify the key issues involved in a negotiation and assess the relative value of the items involved in the exchange. You will also learn about the way that differences in negotiation and communication style affect the way in which you and your client come to an agreement. The video depicts a negotiation between Dr. Derr and Mr. Johnson, and other examples illustrate some of the variety of agreements that can take place between consultant and client.

Communicating about statistics. This is another core communication skill in statistical consulting. When it comes time for you to deliver statistical information to your client, you will be crossing a disciplinary boundary. In order for all of your hard work to have an impact, you will need to present the information in a clear and persuasive way. Typically, statistical consultants deliver statistical information in conversation, by presentation, or in writing. In Chapter 7, you will learn about different theories of learning that can help you be effective in all three of these formats. You will find examples that illustrate how to recognize and adapt to your client's preferences for learning. You will view both positive and negative versions of Dr. Derr presenting her recommendations for the sampling scheme to Mr. Johnson. In the negative version, Dr. Derr does not appear to notice whether or not Mr. Johnson understands her (he doesn't). In the positive version, Dr. Derr adapts her presentation to Mr. Johnson' preferred learning style, with much better results.

Responding to difficult situations. One of the challenges in statistical consulting is the potential for difficult situations to arise. The interdisciplinary nature of the communication, the complexity of the exchanges involved and the unpredictability of real-world projects all contribute to this potential. In Chapter 8 you will find many examples of difficult situations in statistical consulting. You will learn how to recognize a developing problem, discuss it with your client, and work towards a resolution. The video provides an example of a breakdown in the communication between Dr. Derr and Mr. Johnson. You will find out what they do to resolve the breakdown and get their project back on track.

Does this seem like an intimidating set of skills to learn? Nobody is going to master all of them overnight. Even experienced statistical consultants can identify communication skills that they would like to improve. This is part of the reason why statistical consulting can challenge you to grow and develop. Throughout your career, you can broaden your knowledge of statistics and computer applications, increase your experience in problem-solving, fine-tune your work management skills, and improve your communication skills. These efforts will all help improve your effectiveness in statistical consulting.

1.3 How Can Statistical Consultants Improve Their Communication Skills?

As you have seen from the previous section, each chapter of this book concentrates on one set of communication skills. At the beginning of each chapter, you will read about the learning goals of the chapter. To illustrate the skills that are being covered in the chapter, you will find examples drawn from the experiences of actual consultants in a variety of work environments. Some of these examples have successful outcomes and others are not so successful. This will enable you to compare and contrast the features of each example that led to its outcome. Then you can practice your skills with the exercises in the chapter. The exercises include extensions of the examples in the chapter and new examples for you to try.

Communication is a dynamic, interactive process. You can get started by reading about communication skills, but without additional visualization and practice it would be difficult to master these dynamic skills. The video that accompanies this book shows a consultant and a client interacting with each other. You will see positive and negative versions of key segments of the video. This will help you to compare and contrast the outcomes when good and poor communication skills are used.

In addition, you may be able to view or listen to recordings of actual consulting sessions. This can expand your awareness of the many challenges in statistical consulting and the variety of ways to deal with them. You can also learn a lot from the reactions of other statisticians and of clients to these recorded sessions. This book includes suggestions for how an instructor can include recorded consulting sessions in a course or workshop on statistical consulting. These suggestions are included in a section called *Suggestions for Group Discussion* that you will find at the end of each chapter.

Learning about the experiences of other statisticians can also help you improve your communication skills. These may come from the other participants in your class, from your department or unit, or from the world at large. I am confident that you will find statisticians willing to share their experiences and their opinions. Many statisticians would like to improve the overall professional image of statisticians and this is one way that they can help. Some of the exercises in this book ask you to contact statisticians and discuss their experiences and opinions with them. An instructor can find ideas in *Suggestions for Group Discussion* about how to involve experienced statisticians in class activities.

You will ultimately learn a great deal from your clients about the effectiveness of your communication skills. If you are not already consulting, then find some opportunities to get started. The ideas in this book should help you go into a consulting session with an aware and open attitude. If you are able to record a consulting session, you can review it later to critique what went on during the session. You can also ask your client directly for feedback. In all of these efforts, it helps to have someone to coach you. An instructor can find ideas in *Suggestions for Group Discussion* about how to provide these opportunities and the necessary coaching. There are also exercises in the book that ask you to contact clients, and obtain their perspectives about issues. Inviting clients to discuss a consulting topic with the class is a suggestion that can help everyone gain a better understanding of each other.

Before starting out to learn about communication skills, it helps to identify what you would like to accomplish. In the next chapter, you will learn about the qualities of a consulting experience that lead to a satisfied client. You will also be encouraged to identify the attributes that you would like to develop as a statistical consultant. The exercises in Chapter 2 should help you to develop a way to align your goals with those of your client. After working through the chapters and exercises in this book, you can then approach each new consulting experience as an opportunity to improve your skills. The material in the last chapter of the book is geared towards helping you reflect on the outcomes of your experiences: How satisfied were you? How satisfied was the client? Were you able to persuade your client to adopt good statistical practices? What did you learn from the experience that you could apply to the next one? This process will help you integrate the skills from each chapter into a style that is comfortable and effective for you.

Exercise 1.1

Contact a statistician who has had experience in statistical consulting. Find out: (1) his educational background and work experience; (2) what led him to choose a career that involves statistical consulting; (3) what he finds most rewarding and most challenging about statistical consulting; (4) what he would recommend learning from a course in statistical consulting.

1.4 Suggestions for Group Discussion

1. If you are teaching a class of students who are inexperienced in statistical consulting, arrange for them to contact other more experienced statisticians for Exercise 1.1. You can do this in several ways: You can arrange for local statisticians to meet with your students or talk by telephone; you can invite one or more statisticians to come your class; or you can encourage students to contact statisticians via the Internet.

2. If you choose to have students contact statisticians by email, there are several ways this can be done. You can identify consultants in advance who would be willing to respond to your students' emails. You can give students some good starting points (such as the Web address for the American Statistical Association, www.amstat.org) and let them find a statistician on their own. You can also set up an electronic discussion group for all of your students to participate and learn from all of the statisticians who respond.

3. If you are leading a group of more experienced statisticians, you can use Exercise 1.1 as an introductory "ice-breaking" exercise. Participants can break into small groups of 2-3 and interview each other. They can then introduce each other to the class and say something about the background of the person they have interviewed.

1.5 Resources

Ellenberg, J.H. (1999), "Statistics and communication," *AmStat News*, February 1999, 8-9.

2

The Ideal Statistical Consultant and the Satisfied Client

2.1 Introduction

How would you like to be an ideal statistical consultant and have nothing but satisfied clients? As most experienced consultants will tell you, this is more of a lofty goal than a reality. However, before you embark on learning about communication skills, it helps to know what you would like to accomplish with these skills. What qualities does an ideal statistical consultant have? What determines how satisfied a client is with the work that you do? You and your client will both bring your own distinctive qualities to a consulting session. Your training, your experiences, your attitudes and your goals are all built on a foundation of intelligence, personality and beliefs. The way you put these qualities together into an effective consulting style will be as distinctive as your own individuality. This is why we can't develop one single prescription for the qualities of an ideal statistical consultant. The purpose of this chapter is to help you examine your own qualities and develop a profile that is ideal for you. You will also have a chance to look at the statistical consulting experience from the client's perspective. A big challenge in statistical consulting is to make sure that these two perspectives are aligned. Does an "ideal" consultant, as defined from the statistician's perspective, necessarily always have satisfied clients? The two perspectives can actually be quite different! As you will find out, the cause of problems and dissatisfaction in the consulting relationship can often be traced back to this lack of alignment. You can help avoid these problems by making sure that you understand not only your own qualities but also what the client perceives as important to his satisfaction with his work with you.

2.2 Learning Outcomes

- Identify the qualities that you would like to develop as a statistical consultant.
- Compare and contrast the statistical consulting experience from the client's point of view and the consultant's point of view.
- List the attributes of a consulting experience that lead to a satisfied client.
- Identify a strategy for aligning your goals with the client's goals.

2.3 The Statistician's Perspective

Many statisticians have tried to define the qualities of an ideal statistical consultant. In the 1970s, the Section on Statistical Education of the American Statistical Association set up a committee to examine the training needs of statisticians for industry. This committee produced a list of qualities that are important in the general work environment found in industry (ASA, 1980). This list is shown in Exhibit 2.1:

Exhibit 2.1 ***Ideal Industrial Statistician***

- Well trained in theory and practice of statistics
- Effective problem solver
- Good oral and written communication skills
- Can work within the constraints of the real world
- Knows how to use computers to solve problems
- Is familiar with the statistical literature
- Understands the realities of statistical practice
- Has a pleasing personality and is able to work with others
- Gets highly involved in the solution of company problems
- Is able to extend and develop statistical methodology
- Can adapt quickly to new problems and challenges
- Produces high-quality work in a timely fashion

A statistician with all of the qualities listed in Exhibit 2.1 sounds ideal indeed! It can be intimidating to inspect a list like this and then evaluate one's own qualities and achievements against it. Boen and Zahn did just this in Chapter 4 of *The Human Side of Statistical Consulting* (1982). They assigned themselves letter grades ranging from A's to C's on each quality on the list. They pointed out that even very experienced and able consultants would not have a perfect "A" grade for every quality. They concluded that a list like this is helpful as long as it is inspiring rather than discouraging. It is best to think of these ideals as goals that can help guide your progress.

More recent descriptions emphasize a statistician's broad participation in the team, the project, and the workplace. Today's ideal statistical consultant appears to get fully involved to solve problems that have more than just a statistical dimension. For example, the National Research Council sponsored a forum in the 1990s on the statistical needs of industry and government. In this forum, Jon Kettenring made this comment about the needs of industry:

> Along with the uncertainties are immensely challenging and exciting problems that need serious work, and much of this work is statistical in nature. Examples close to home include network reliability, software quality, and dealing with massive data sets.

> Generally speaking, there is a need to transform data into information … and intelligence that cuts across much of modern industrial life. The major problems are usually of such complexity that progress is best made when cross-organizational interdisciplinary teams tackle them. Since many of these problems cry out for statistical thinking, it is natural for statisticians to be full partners on the teams. However, to be successful, statisticians need not only to bring their statistical expertise to the table but also to integrate themselves effectively with the rest of the team. Ultimately, the only thing that really matters from the industrial perspective is not the statistics per se but the impact of the total effort on the problem at hand.[1]

In the same forum, N. Phillip Ross discussed the statistical needs of government. He wrote:

> The most important thing I would like to see is people emerging from graduate school understanding that they are going to play on a team, and knowing how to communicate in that team setting. That is very difficult for people, especially those majoring in mathematics and statistics.2

He then related a story from his experience (Example 2.1):

Example 2.1

When I went to work, initially in the private sector and then in the government, I remember talking with managers in a roundtable on an approach about which I said, "If you take that sample size, then you cannot infer." The person managing this roundtable said, "Hold it, speak English." Prior to that, I had not known that using "sample" and "infer" would pose a problem of understanding for some people. ... One of the critical things I have learned through experience in the federal government is to be careful about language, to be careful about jargon, and to become a member of a team that solves a problem as opposed to showing off my technical mastery, so to speak. Currently, statisticians who are hired straight out of the university must learn that on the job. Some do and some do not. As indicated earlier for industry, we at EPA [the Environmental Protection Agency] also can no longer spend the time on that to acquire training on the job. ... Having experience in teamwork is very important, especially with respect to communications. We want people who can communicate easily and across disciplines, and who will have the other team members understand what statisticians are trying to tell them.[3]

---

[1] From "What Industrty Needs" by J. Kettenring, p. 2. Copyright © 1995, American Statistical Assocaition.
[2] From "What the Government Needs" by N. Philip Ross, p. 7. Copyright © 1995, American Statistical Association.
[3] From "What the Government Needs" by N. Philip Ross, p. 7. Copyright © 1995, American Statistical Association. Reprinted with permission.

Gerry Hahn and Roger Hoerl provide a similar perspective in their 1998 article on key challenges for statisticians in business and industry. Here is what they wrote about the contemporary role of the statistician:

> More than ever before, business and industry require statisticians who have the insight to get at the real root causes of issues, both technical and non-technical; general problem-solving and scientific thinking skills; a broad base of statistical and subject-material knowledge; the ability to learn quickly what they do not know; the skill to adapt such knowledge to the problem at hand; the confidence to work effectively with others in teams; the willingness to balance thoroughness with getting the job done on time; the stomach to work in a dynamic, high-pressure environment and to put in long hours when needed; outstanding communications and training skills; and the enthusiasm (and immodesty) to sell themselves and their ideas.... In short, statistical knowledge is a necessary, but far from sufficient, condition for a successful career in business or industry. … Moreover, the statistician in business or industry must become a deeply immersed and proactively committed team member - rather than a consultant whose advice is sought only on occasion ... Being a contributing team member requires a broad approach to problem-solving -- that is, getting out of one's technical "box" to learn about other disciplines and often performing tasks that do not directly relate to one's narrow specialty.[4]

A 1998 commentary by Margaret Nemeth also reinforces this view of the statistician as a fully involved member of a team or organization:

> With the changing and evolving business culture along with the team-based flat organization, individuals no longer 'move up the career ladder.' At Monsanto everyone is expected to contribute in the team environment and 'rank' is no longer used. In fact, we are not allowed to have any rank designation in the title we choose for our job description. Everyone is responsible for his/her own growth, both professional and personal. We are expected to coach each other in areas where we need to grow or improve. We get out of the job what we put into it, and we are expected to maximize our potential.[5]

The message from these and other statisticians who have had experience in business, government and industry is that today's ideal statistician needs to know how to work collaboratively with others. They would like statisticians to consider not just the narrow statistical aspects of a problem but also to contribute to the broader objectives of a project and goals of an organization.

---

[4] From "Key challenges for statisticians in business and industry" by Gerry Hahn and Roger Hoerl, p. 197. Copyright © 1998, American Statistical Association. Reprinted with permission.

[5] From "Discussion" by Margaret A. Nemeth, p. 206. Copyright © 1998, American Statistical Association. Reprinted with permission.

Exercise 2.1

Choose a field of statistical consulting that interests you. (For example, you might choose the biopharmaceutical industry, engineering or agriculture). Find out what qualities are important to statisticians working in that field. You may be able to locate publications that address this topic. Use the references at the end of this chapter, listed under the heading "Articles about qualities of statistical consultants" as a starting point. If you can, contact a statistician practicing in this field and obtain her perspective. Develop a list of "ideal qualities" for statisticians working in this field.

Exercise 2.2

Examine the qualities of an ideal statistical consultant that you have read about in this section. Try to group and characterize them. For example, it might be helpful to consider which ones are based on personality, attitudes, intelligence, education, experience and so on. Then select three qualities that are quite different from each other. For each one, suggest ways that a person could acquire this quality or increase their level of skill or competence. Are there any qualities you feel that a person could not change by some means? If so, identify these and explain why.

Exercise 2.3

Identify qualities that you would like to have as an ideal statistical consultant. Privately, assess yourself on these qualities. Choose one that you feel would be relatively easy for you to attain or improve and develop a plan for doing this.

2.4 Multiple Modes of Intelligence

Today's ideal statistical consultant has qualities that originate in knowledge, skills, experience, attitudes, work habits, personality, motivation and behavior. This is one reason why statistical consulting is such an interesting profession. As a statistical consultant, you will be challenged to function well in nearly every mode of intelligence. The theory of multiple modes of intelligence is helpful in illustrating the demands on a statistical consultant. This theory was developed by Howard Gardner, a cognitive psychologist, and has had widespread acceptance in educational psychology. Following this theory, "intelligence" is defined as the multiple ways by which a person understands and learns about his environment, and solves problems and creates products or services that are valued in his environment. There are usually seven modes of intelligence identified in this way:

Verbal / Linguistic intelligence includes the verbal abilities of speaking and writing. It encompasses not only language but also the abstract reasoning, symbolic thinking and conceptual patterning that form the basis for language. Someone with a well-developed verbal /

linguistic intelligence may have an extensive vocabulary, learn languages easily, and enjoy reading, writing and telling stories. This mode of intelligence is important in statistical consulting because of the need to transmit statistical information through language.

Logical / Mathematical intelligence includes both inductive and deductive reasoning. If you have chosen a statistical career, chances are that you enjoy working with abstract symbols and recognizing patterns and relationships between different pieces of information. An essential part of research is the ability to observe objectively, and then by inductive reasoning formulate hypotheses and draw conclusions from data. Statisticians also make use of deductive reasoning to apply abstract rules and principles to specific problems and applications.

Visual / Spatial intelligence includes the ability to form images and pictures in the mind, and solve problems in spatial dimensions. Engineers, artists, designers, mapmakers, and chess players all make use of visual / spatial intelligence in their pursuits. In statistics this intelligence is important in producing statistical graphics and in working with abstract spaces that have multiple dimensions.

Interpersonal intelligence includes the ability to establish relationships with other people. Someone with well-developed interpersonal intelligence will be able to communicate well with others, and readily sense another person's moods and intentions. Leadership, empathy, and social organization all require interpersonal intelligence. This mode of intelligence is essential to the many interpersonal interactions in statistical consulting.

Intrapersonal intelligence includes self-knowledge and transcendent reflection. Someone with a well-developed level of intrapersonal intelligence may prefer to spend time alone in introspective thinking. Philosophers, spiritual and religious thinkers make use of intrapersonal intelligence. Statisticians may also make use of this type of intelligence in developing abstract theory. In this chapter you are being asked to reflect about yourself, your qualities and goals for yourself as a statistical consultant.

Body / Kinesthetic intelligence is expressed through the movement of the body. Athletes, dancers, and actors all make use of this intelligence. Many people who have a strong kinesthetic component to their intelligence learn best by actively handling and manipulating objects. Statistical consultants need to keep this in mind when they consider how to communicate statistical knowledge to someone who prefers a "hands-on" method of learning.

Musical / Rhythmical intelligence includes abilities such as recognizing and remembering musical patterns and rhythms, playing a musical instrument, singing, and writing music. It is difficult to see how this mode of intelligence is involved in statistical consulting, but a future reader may be able to make a connection.

You and your clients probably have different strengths in these seven modes of intelligence. After all, you are in different professions. Gardner's indirect contribution to statistical consulting was to foster an appreciation of the diverse ways that intelligence could be defined and expressed in different people. This is consistent with the consultant's need to work collaboratively with people in other disciplines. In a good working relationship these differences become strengths rather than weaknesses. By bringing more than one way of looking at and solving problems to a project, you and the other members of a team can pool your strengths and reach more robust solutions more efficiently than if each of you were working separately. However, this ideal merging of complementary strengths does not always happen. Sometimes other factors interfere with the smooth functioning of a team. For example, differences in personality between you and your client can create difficulties. Boen and Zahn's book *The Human Side of Statistical Consulting* (1982) includes some insightful descriptions of the personalities of consultants and clients and useful vignettes of what can happen when people with incompatible personalities must work together to solve problems. In this book you will not find a focus on personality. Instead, you will read about how differences in expectation, perspective and style between you and your client that, without due attention, can interfere with your effectiveness.

Exercise 2.4

Return to the qualities of an ideal consultant that you had characterized in Exercise 2.2. Now assign them to different modes of intelligence. Interpret your findings.

Exercise 2.5

Privately, return to the self-assessment you did in Exercise 2.3. Identify modes of intelligence where you have special strengths and others where you could usefully increase your development in order to improve your skills as a statistical consultant.

Exercise 2.6

Can you identify one of your clients whose strengths across these modes of intelligence is fairly similar to yours? Can you think of a client whose strengths are markedly different from yours? How do these similarities and differences affect the working relationship that you have with each one? If you haven't had much consulting experience yet, then think of others with whom you have worked in another capacity.

2.5 The Client's Perspective

There is less information in the statistics literature about what statistical consulting is like from the client's point of view. Statisticians appear to have neglected to gather data on their performance in a very systematic way from the most direct source of information, their clients! Therefore the information available at this point is somewhat anecdotal, but it still provides thought-provoking reactions and a starting point for your own investigation.

Gathering feedback from clients is one way to get a better picture about what features of the consulting experience lead to satisfaction and what leads to dissatisfaction. Here are some comments taken from a survey of customers who visited a university consulting unit. The questionnaire asked clients to rate their level of satisfaction with different aspects of the consulting experience. Space was also provided on the questionnaire for open comments. Some were very satisfied (Exhibit 2.2) and others very dissatisfied (Exhibit 2.3) with the consulting they received.

Exhibit 2.2 ***The Satisfied Client***

1. "I was very impressed with the effort the consultant and assistants made to understand the problem...."

2. "[Consultant] was very accommodating, professional and very accessible. The documentation was very readable, and the technical material was presented in a way that was easy to understand as well as implement. Overall, my interactions with [Consultant] were very satisfying."

3. "I enjoyed working with [Consultant]. She clearly enjoys her work, and is one of those rare people who is also capable of putting it into words that non-statisticians can understand. Thank you."

4. "I was very pleased with the assistance I received. The consultant was very professional in her approach to the problem."

5. "The document [Consultant] prepared was extremely helpful - clear, easy for me to understand and it addressed everything I was concerned about."

6. "Thanks very much for your help! You gave me exactly the lead that I needed and I never would have found it on my own. I bought a copy of the book she sent a chapter from - it will be very useful in future work."

7. "[Consultant] truly seemed concerned, that he truly desired to aid in my problem. I believe that his good attitude personifies what is required in consulting practices."

These clients had very strong reactions to their consulting experiences! By comparing and contrasting the positive comments with the negative ones, you can develop some ideas about what features of the consulting experience are important to a client's level of satisfaction. You may also find it helpful to learn about what experts have written in the general field of customer satisfaction. A useful reference is *Measuring Customer Satisfaction: Development and Use of Questionnaires* by Bob Hayes (1992). In this book, Hayes identified five dimensions of quality

Exhibit 2.3 ***The Dissatisfied Client***

8. "I was able to follow up on most of the recommendations - with a lot of reading. However, I still feel confused about some aspects of the results and I feel as though I have no where to turn for answers. I really need to have someone to ask questions of in addition to the two scheduled meetings. While I have 'results' I do not know if they are accurate."

9. "The time frame in waiting for [Consultant]'s recommendations was longer than I expected. Otherwise, I think it was a learning experience for both of us."

10. "This experience was unacceptable to me as a researcher.... [Consultant] was consistently late to meetings - some time not showing up at all. Also after last contact with him he did not return any e-mail and dropped out of sight?!"

11. "Since you asked, I will tell you that I received poor advice. I did not know it at the time, but when I went to [get help from someone else, I] ended up changing the model significantly to make it right. I waited 3 weeks for feedback from [Consultant] and since it turned to be wrong, I wasted the whole month of March trying to use the [Consulting Services]."

12. "[Consultant] did understand my questions, but had a very difficult time justifying the method used to obtain the answers [I was looking for]."

that determine the level of a customer's satisfaction with a product and/or service. These are quoted in Exhibit 2.4. These dimensions are relevant to statistical consulting, and they are qualities that you can improve through your own efforts. Let us consider each dimension:

Availability of support. Your client would like to be able to reach you. She would also like for you to provide the amount of support that the two of you had agreed upon. The consultant who inspired comment #10 in Exhibit 2.3 had not showed up for some meetings and had not returned the client's emails. From the client's comments it appears that the consultant had abandoned the client. Of course we don't know the whole story behind this comment, but we can speculate that the consultant was not clear with the client about how much support he intended to provide. He and his client probably did not talk about how to reach each other, or about the circumstances of future communication. You can avoid this type of dissatisfaction by developing clear consulting agreements with your client. Does this mean you have to be constantly on call, 24 hours a day, 7 days a week? Of course not! Being available to your client means that you and your client have discussed and agreed upon how and under what circumstances you will contact each other and what support you will provide.

Exhibit 2.4 ***Five dimensions of quality which determine a customer's level of satisfaction*** [6]

1.	*Availability of support*: the degree to which the customer can contact the provider.
2.	*Responsiveness of support:* the degree to which the provider reacts promptly to the customer.
3.	*Timeliness of support:* the degree to which the job is accomplished within the customer's stated time frame and/or within the negotiated time frame.
4.	*Completeness of support:* the degree to which the total job is finished.
5.	*Pleasantness of support:* the degree to which the provider uses suitable professional behavior and manners while working with the customer.

Responsiveness of support. Your client would like you to respond promptly to her requests. Does that mean you have to drop everything every time a client asks you to do something? Of course not! If you did do that, pretty soon nobody would be satisfied and you would be a burned out husk of a consultant. A very big part of being responsive is being able to communicate clearly to your client that you understand what it is she is asking from you. Then you need to be clear about whether or not you plan to work with this client. If you do, then you and your client need to agree on what will be done by whom and by when.

Timeliness of support. Timeliness plays an important part in customer satisfaction. Several dissatisfied clients who contributed comments to Exhibit 2.3 complained that the help they received took too long, or came too late. This is why you and your client should discuss and agree on deadlines and priorities. It helps then to stay in contact with the client so that you can update the deadlines as your and your client's circumstances develop.

Completeness of support. Several of the dissatisfied clients in Exhibit 2.3 complained that they needed more help that wasn't available, or they didn't understand the advice that was given to them. You can avoid this problem by having a clear understanding with your client about what your tasks are. In statistical consulting, a client can often feel that support is incomplete if she doesn't understand your statistical work and is unable to apply what you have done to her problem or project. Your skill in explaining your statistical work will help a client feel that your work has come to closure satisfactorily.

[6] From *Measuring Customer Satisfaction: Development and Use of Questionnaires* by Bob E. Hayes, p. 8. Copyright © 1992, ASQC Press. Reprinted with permission.

Pleasantness of support. Being able to convey your interest in the project and willingness to work with the client will have an impact on your client's level of satisfaction. Many of the positive comments in Exhibit 2.2 took note of the consultant's interest and apparent enjoyment of the assignment. Other comments praised the consultant's level of professional behavior. Does this mean that you have to like every single thing that you do for every client? Of course not! Many statistical consultants are in work environments that don't permit them full choice about who they will be working with and what they will be doing. Professional behavior is all about acting with courtesy and respect, even under trying circumstances.

What about the quality and accuracy of your statistical work? Shouldn't that be a key dimension of customer satisfaction? Maybe it should be, but keep in mind that it is difficult for a non-statistician to evaluate your statistical work directly. In some instances, a client may be able to send your work out for an independent statistical review. However, in most consulting situations the client relies on indirect measures of the quality of your statistical work. The reputation of where you work, your academic degrees, your resume and the recommendations of others are all indirect measures that your client may use to judge the quality of your statistical work.

Statisticians can become very frustrated when they know they have produced excellent statistical work for a client who reacts with dissatisfaction. The client may perceive this work as incomplete because they do not understand it or are unable to apply it to their problem. Margaret Nemeth provides an example of this situation in Example 2.2:

> **Example 2.2**
>
> Just yesterday I met with one of our analytical chemists from a California subsidiary [of Monsanto]. This location has 700 researchers, none of which is a statistician. He was here in St. Louis to request my help in optimizing a new process and in validating an analytical method. The location had contracted with a local university for statistical consulting but it was a disaster because 'the statistician gave us a lot of theory but could not apply it to our problems.' Thus, the problems remained unsolved. What impression do we as statisticians leave behind when we walk out of a client's office? Why are we continually surprised by people's attitudes about statistics and statisticians in general? What should we be doing to change our practices so that we in turn change these attitudes? [7]

I asked Dr. Nemeth to give me her assessment of the customer satisfaction issues in this anecdote. Here is what she wrote:

> The customer in Example 2.2 was dissatisfied because he had paid for a service with neither short-term nor long-term value. First, he received a lot of statistical information but with no insight as to how it applied to his problems. Second, he

---

[7] From "Discussion" by Margaret A. Nemeth, p. 206. Copyright © 1998, American Statistical Association. Reprinted with permission.

felt that he had received no help on how to solve the current problems. Since the statistical consultant whom he had contacted was from a university, the consultant may not have been familiar with the types of problems (optimization and method validation) being addressed. Optimization uses response surface methodology and method validation uses variance component estimation along with the concept of nested designs. These topics are not always addressed in statistical curricula. The researcher also felt that the statistician could not communicate on his level. It takes effort to explain statistical concepts on a level that a non-statistician can understand. However, communication is always two-way and the researcher needs to be able to communicate the chemistry concepts on a level that the statistician can understand. I do believe that it is easier for the statistician to facilitate the flow of information from the chemist (researcher) to the statistician because he/she normally asks many questions so that he/she can thoroughly understand the "process". It is more difficult for the researcher to obtain the statistical information because of the nature of the statistical discipline and the fact that the researcher just does not know what questions to ask if he/she does not understand what the statistician is saying. Thus, it is very crucial that the statistician speak on a level that can be understood by a non-statistician.[8]

The statistician in Example 2.2 was not able to communicate his statistical knowledge in a way that the client could apply it to his problem. You can see the level of dissatisfaction that this generated. The project seemed incomplete from the client's perspective.

When a statistician is able to bridge the gap between disciplines, she can really have an impact. In Example 2.3, also provided by Nemeth (1998), the statistician was able to persuade her clients to adopt the methods of statistical process control:

Example 2.3

When I introduced SPC [statistical process control] into our Brazil manufacturing facility approximately four years ago, I tied SPC to experimental design and process improvement. I spent two weeks at the plant becoming familiar with the different processes and discussing both SPC and experimental design. I had been told that the plant had been trying to introduce quality control charting for approximately six years. They brought several consultants into the plant, but after the consultants left the issue died and the charts were not implemented. I returned to the plant with some trepidation 18 months later in a follow-up visit. I was pleasantly surprised to find control charts up and running on almost all the processes. Moreover, experimental design techniques were

---

[8] Margaret A. Nemeth, personal communication, 1998. Reprinted with permission.

> being used for process improvement. And just recently we started control charting as part of our round robin testing for our glyphosate analytical assay.[9]

Example 2.3 is a "success story." Dr. Nemeth was able to persuade her clients to adopt good statistical practices, and the clients were satisfied with her work. She spent time at the plant and appeared to get to know the people and their problems well enough that she was able to gain acceptance for the implementation of statistical process control. Because she was effective, she was able to have an impact on industrial processes that were important to her organization. I asked Dr. Nemeth to comment on the satisfactory outcomes of this example Here is what she wrote:

> The success of the project in Example 2.3 again hinged on communication and the ability to communicate statistics in a clear and concise fashion to non-statisticians. Another important reason for its success was the fact that the statistician had convinced the quality control supervisor for the plant that process improvement and quality control charting would improve the quality of the products and, more importantly, reduce customer complaints and save money. As an aside, communication includes not only the skill of discussing statistical concepts on a level that a non-statistician can understand but also the ability to: be friendly, be a good listener, show interest in the client and his/her research problem, ask questions, be personable and most importantly never ever make a client feel inferior[10].

Communication is the key to all five dimensions of customer satisfaction. When you and your client clearly understand and agree upon the circumstances of your work together, this will promote satisfaction. You will appear to be responsive and available with this level of clarity. Behaving professionally and staying in touch about deadlines will also promote satisfaction. Finally, being able to communicate your statistical messages clearly so that a client can understand and implement them provides closure to a project. This will promote satisfaction that the project is complete.

Exercise 2.7

Read through the comments from dissatisfied and satisfied clients in Exhibits 2.2 and 2.3. For each comment, identify the dimensions of customer satisfaction from Exhibit 2.4 that are involved.

---

[9] From "Discussion" by Margaret A. Nemeth, p. 206. Copyright © 1998, American Statistical Association. Reprinted with permission.

[10] Margaret A. Nemeth, personal communication, 1998. Reprinted with permission.

Exercise 2.8

Research in the area of customer satisfaction has suggested that customers express satisfaction when their expectations are either met or exceeded. They express dissatisfaction when their expectations are not met (Swan, et al. 1984). Choose two comments from the satisfied customers (Exhibit 2.2). Describe what expectations were either met or exceeded. Then choose two comments from the dissatisfied customers (Exhibit 2.3). What were the expectations that were left unmet? Feel free to embellish the "stories" behind these comments as needed to illustrate your point of view.

Exercise 2.9

Work with the two comments you selected from dissatisfied customers in Exercise 2.8. How would you go about improving things so that future customers would not be dissatisfied in this way? Feel free to embellish the "stories" behind these comments.

Exercise 2.10

Interview a client. Find out what aspects of a statistical consulting experience are most important in determining here level of satisfaction. Summarize the interview. Relate her comments to the dimensions of customer satisfaction we covered in Exhibit 2.4.

2.6 Aligning Expectations

Clear communication between you and your client will help you both feel satisfied with the outcome of your work together. When you and your client have different expectations about the arrangements you have made, either one or both of you may end up feeling dissatisfied. In order to prevent this dissatisfaction, what topics should you and your client discuss and agree upon? In statistical consulting, there are ten issues that represent special vulnerabilities in the client-consultant relationship. Lack of clarity about these topics is often at the root of misunderstandings, unmet expectations and dissatisfaction. Although you probably won't have to discuss all ten issues with each client, it is good to keep them all in mind when you are setting up a new consulting agreement. These ten issues are as follows:

Issue #1: What is your role?

Your role in a consulting project is defined by three components: (1) your responsibilities; (2) the level of authority that you have, and (3) the way that you participate in making decisions. This can vary a lot from project to project. You can become dissatisfied when this role has not been well defined. Another source of difficulty occurs when the three components of your role do not correspond well with each other; for example when you carry a great deal of responsibility but do not have much authority or much access to the decision-makers.

Issue #2: What are the roles of others on the project?

Closely tied to your role is the part that others will play in the project. It is important to find out the division of responsibilities and authority in a project and what the decision-making process is. As with Issue #1, problems arise when these are not well defined or when they do not correspond sensibly with each other.

Issue #3: How will communications be maintained?

Projects are dynamic, and when you stay in regular communication with your clients you are most likely to keep abreast of the changes. This will enable you to expend your efforts in the most efficient way. Many different kinds of problems can arise when communications are inadequate among the members of a project team.

Issue #4: What are the "deliverables"?

Let us define "deliverable" in the most general possible sense to mean whatever you have agreed to provide for the client or the project team. If you spend a lot of time and expertise in producing deliverables that the client does not perceive of as relevant, then neither you nor the client is likely to be very satisfied.

Issue #5: What are the deadlines?

Timeliness is a key dimension of customer satisfaction. Sometimes a deliverable that is one day late has no value. You should know what a project's deadlines are and how flexible they are, so that you can evaluate whether or not you can meet them.

Issue #6: How will you be compensated for your participation?

Let us also define "compensation" in the most general possible sense to mean whatever the client has agreed to provide to you in exchange for your work. Dissatisfaction can arise when you and your client have different interpretations of how you will be compensated for your efforts.

Issue #7: What are acceptable statistical practices?

You and your client or project team should be clear about what statistical practices are acceptable for a project. Problems can arise if these agreements are not made in advance of summarizing and analyzing data from a project.

Issue #8: What are the ownership rights?

It is natural for you to feel a sense of ownership of a project after spending a lot of time and brainpower on it. However, it is a really good idea to establish with your client the circumstances under which you may make future use of information from the project or make this information known to others.

Issue #9: What stipulations are there for security and confidentiality?

You are responsible for maintaining the security and confidentiality of the information that you have received from your client. You should make sure that you and your client both understand what the expectations are and what the capabilities and limitations are of your workplace.

Issue #10: When is your participation finished?

Sometimes it seems as if a project is never finished! You may want to close off your participation in a project and direct your efforts to other activities. The client may want to continue to explore the data or extend the project into other promising directions. These two different perspectives can lead to dissatisfaction at the end and may overshadow your earlier good work. For this reason it is a good idea to agree on the circumstances under which your participation in a project has finished. At that point you can have the option to continue under a new agreement.

There seem to be a lot of vulnerabilities in the client-consultant interaction where dissatisfaction can arise! This is part of the challenge of statistical consulting. The rewards are just as great or even greater when you and your client establish a satisfying working relationship. You and your project team will also feel this satisfaction when you work together to develop a technically sound solution to a problem. Many of the potential sources of dissatisfaction discussed in this chapter can be avoided by timely and clear communication. Once you and your client are aligned on your expectations, you can focus your energies on providing your high quality statistical efforts towards a satisfactory outcome.

Exercise 2.11

Statisticians have written articles about their preferred roles in different job environments. (1) Read at least one article on this topic. There are listings at the end of this chapter to get you started under the heading *Resources about the role of statistical consultants*. (2) Describe the preferred role for the statistician. (3) Identify any dissatisfaction with this role that the author has described. (4) Frame any dissatisfaction you have read about in terms of divergent expectations on the part of the statistician and the client. (5) If you were the statistician in this situation, how would you go about identifying these divergent expectations as early as possible in your work with your client? (6) How would you attempt to align your and your client's expectations of your role in order to achieve better satisfaction?

Exercise 2.12

Choose a work environment for statistical consulting that interests you. Find some examples of explicit policies of ownership rights, security and confidentiality, and acceptable statistical practices that are used in this work environment.

2.7 Suggestions for Group Discussion

1. Class discussion is an essential part of this chapter. Emphasize that there are no right or wrong answers! Encourage the participants to analyze and draw generalizations from their findings, and to make comparisons between the perspectives of statisticians and clients.

2. Consider setting up an electronic discussion group involving your class, some statisticians and clients for some of the exercises in this chapter. Try to maximize the diversity of statisticians and clients who contribute information to the class as a whole.

3. You can arrange for clients and experienced statisticians to come talk to the class. Consider having clients and statisticians comment on the descriptions of the ideal statistician, the satisfied customer and the dissatisfied customer. Try to invite clients and statisticians who represent different work environments and experiences.

4. A class can be subdivided into small working groups to do some of the exercises. You can assign different items within an exercise to different groups and have them report back to the class.

5. Exercises 2.3 and 2.5 ask for a self-assessment of a student's strengths and weaknesses in different qualities of a statistical consultant. This should be a private and optional exercise. Consider making yourself available for private conversations about a student's self-assessment.

2.8 Resources

Articles about qualities of statistical consultants:

ASA Section on Statistical Education Committee on Training of Statisticians for Industry, (1980), "Preparing statisticians for careers in industry," *The American Statistician,* 34, 65-75.

ASA Section on Statistical Education Committee on Training of Statisticians for Government, (1982), "Preparing statisticians for careers in the federal government," *The American Statistician,* 36, 69-81.

Boen, J.R. and Zahn, D.A. (1982), *The Human Side of Statistical Consulting*, Belmont, CA: Lifetime Learning Publications.

Hahn, G. and Hoerl, R. (1998), "Key challenges for statisticians in business and industry," *Technometrics,* 40, 195-200.

Hoerl, R.W., Hooper, J.H., Jacobs, P.J., and Lucas, J.M., (1993), "Skills for industrial statisticians to survive and prosper in the emerging quality environment," *The American Statistician,* 47, 280-292.

Hogg, R.V. (1985), "Statistical education for engineers: An initial task force report," *The American Statistician,*39, 168-175.

Kettenring, J. (1995), "What industry needs," *The American Statistician,* 49, 2-4.

Kettenring, J. (1997), "Message to students: Will you get a job in industry?" *AmStat News,* 240, 9-10.

Marquardt, D.W. (1981), "Criteria for evaluating the performance of statistical consultants in industry," *The American Statistician, 35, 216-219.*

Nemeth, M.A. (1998), "Discussion," *Technometrics,* 40, 206-207.

Ross, N. P. (1995), "What government needs," *The American Statistician,* 49, 7-9.

Resources about the role of the statistical consultant:

Boroto, D. R. and Zahn, D. A. (1989), "Promoting statistics: On becoming valued and utilized." *The American Statistician,* 43, 71-72.

Greenfield, A.A. (1979), "Statisticians in industrial research: The role and training of the industrial consultant," *The Statistician,* 28, 71-82.

Gross, I.D.J. (1974), "The role of the statistician: Scientist or shoe clerk," *The American Statistician,* 28, 126-127.

Hunter, W.G. (1981), "The practice of statistics. The real world is an idea whose time has come," *The American Statistician,* 35, 72-76.

Marquardt, D.W. (1979), "Statistical consulting in industry," *The American Statistician,* 33, 102-107.

Williford, W.O., Krol, W.F., Bingham, S.F., Collins, J.F., and Weiss, D.G. (1995), "The multicenter clinical trials coordinating center statistician: 'More than just a consultant', " *The American Statistician,* 49, 221-225.

Resources about types of intelligence:

McPhee, D. (1996), *Limitless Learning: Making Powerful Learning an Everyday Event.* Tucson, AZ: Zephyr Press.

Gardner, H. (1983), *Frames of Mind: The Theory of Multiple Intelligences.* New York: Basic.

Lazear, D.G. (1990), Seven *Ways of Knowing: Teaching for Multiple Intelligences.* Australia: Skylight Publishing, Inc.

Resources about customer satisfaction:

Hayes, B.E. (1992), *Measuring Customer Satisfaction: Development and Use of Questionnaires.* Milwaukee, WI: ASQC Quality Press.

Swan, E., Sawyer, J., VanMatre, J., and McGee, G. (1989), "Deepening the understanding of hospital patient satisfaction: Fulfillment and equity effects," *Journal of Health Care Marketing* 5:7-18.

3

NON-VERBAL COMMUNICATION

3.1 Introduction

Have you ever stopped to think about all of the ways that you communicate without using words? Your facial expression, your eye contact, your tone of voice, the position and motion of your body, and the layout of your meeting place are all examples of non-verbal communication. It is true that you will be conveying statistical information to your client through words and pictures. However, non-verbal communication plays a very important part in statistical consulting. In fact, before you have had a chance to make any statistical comments at all, these non-verbal elements of communication have already conveyed impressions to your client about you such as *welcoming, busy, interested, not interested, intimidating, nervous, professional, disrespectful*, *competent,* and so on. As you learned from Chapter 2, your client's level of satisfaction is driven in part by how available, responsive and pleasant she perceives you to be. Your non-verbal communication is the first way that you will convey these qualities. Any tension created by poor non-verbal skills will interfere with your statistical communication. This chapter will show you how to convey your good intentions to your client through non-verbal communication.

When two strangers meet, their primary concern is to get to know each other well enough to reduce the uncertainty about each other's behavior (Ting-Toomey and Korzenny, 1991). When people from different cultures meet for the first time, they have some special barriers to overcome. In general, these two people will tend to see each other as less similar to themselves than they would if they both came from the same culture. They will tend to seek less information from each other, disclose less about themselves, and have shorter meetings in comparison with two people from the same culture. However, once the barriers to inter-cultural communications are overcome, culture does not appear to be a major factor in subsequent interactions. This is why it is important for you to pay attention to your first contact with a client who is from a different culture. In this chapter you will explore the interpretations that people from different cultures may give to non-verbal communication, and develop strategies to identify and adapt to these differences.

3.2 Learning Outcomes

- Define how you will use non-verbal communication to convey a good impression to clients in a meeting.
- Identify how you will discover and adapt to cultural differences between you and your client in norms for nonverbal communication.
- Compare and contrast the opportunities and limitations for creating a good impression in different media typically used in statistical consulting.

3.3 First Impressions Are Influenced by Non-Verbal Communication

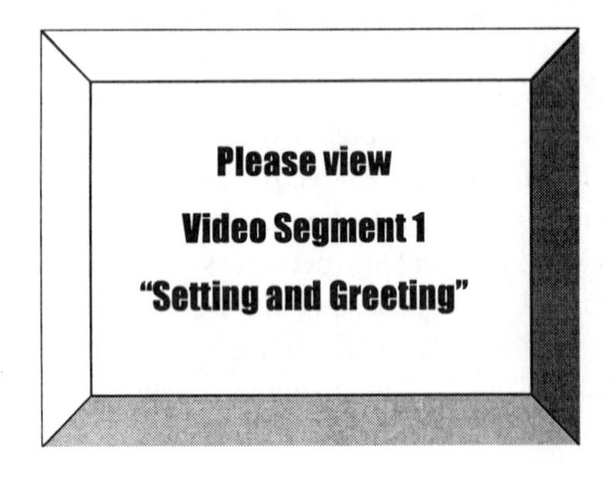

In a meeting, most clients have created a first impression of you before you make a single statistical comment. The setting of the meeting room and your greeting convey the first messages about what you want the client to think of you. This is easier to illustrate them in a more dynamic visual medium than on the printed page. If you are able to view segment 1 of the video before continuing with this chapter, please do so now. You will see one version of the greeting between Dr. Derr and Mr. Johnson. As you watch this segment, take note of the physical layout of the room as well as the greeting. If it helps you to focus on the non-verbal aspects of the greeting, turn the audio off. Once you have finished viewing segment 1, reflect about the overall first impression that the setting of the room and the greeting have given to Mr. Johnson.

As you might have guessed, segment 1 was a negative version of a greeting. There were several physical barriers to the greeting, including books piled high on Dr. Derr's desk and on the chair that Mr. Johnson was supposed to sit in. Dr. Derr did not leave her desk, and she did not make much eye contact or smile at Mr. Johnson. None of the physical and non-verbal cues gave Mr. Johnson much encouragement or made him feel welcome.

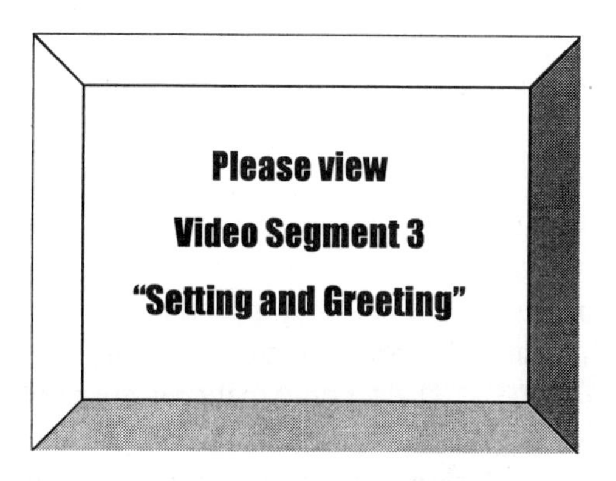

Now, take a look at the positive version of this first contact, which is depicted in segment 3 of the video. Again, turn the audio off if you'd like to focus better on the non-verbal aspects of this segment. Notice the features of the setting and the greeting that make this a more positive greeting. Dr. Derr is certainly friendlier; she meets Mr. Johnson at the door, smiles and shakes his hand. The meeting room is more welcoming also; there is a round table with two chairs, and no physical barriers to the discussion.

The setting. The layout of a meeting room communicates messages about you to your client. Research has shown that lighting, color, and the type and arrangement of office furniture all affect people's perceptions and ability to work on different types of tasks. In the negative version of the first meeting on video, Dr. Derr speaks across her desk to Mr. Johnson. This can put the consultant in a dominant position with respect to the client. Implied in the setting is that the consultant is of higher status than the client. In the positive version, Dr. Derr and Mr. Johnson are seated together at a round table, positioned to cooperate on a task. This setting implies that the consultant and the client are of equal status, and can help to establish a collaborative relationship.

The greeting. The way you and your client greet each other can certainly influence the comfort of the forthcoming discussion. There is no quicker way to get a meeting started off badly than to expect one greeting and get a different one. If this happens to you, it is easy to have a quick emotional reaction such as "How rude!" In segment 1, Dr. Derr waited for Mr. Johnson to come into her office and sit in the chair. She said hello, but she did not shake hands. This behavior can convey that the consultant is not very interested in the client, and just wants to complete the meeting as soon as possible. It is okay for you to wish to speak only briefly with a client. However, you can convey your limits in a more respectful manner.

In segment 3, Dr. Derr greeted Mr. Johnson at the door, made eye contact, smiled, shook hands, spoke, and ushered him to the round table. This greeting conveys respect, and sets the stage for a productive meeting. Both parties in this segment used the convention of a handshake for a greeting. However, there are other forms of greeting used in other cultures. Bowing, bringing the hands together, slapping hands, and kissing cheeks are all examples of other styles of greeting. Some people are distinctly uncomfortable with the touch involved in shaking hands, kissing cheeks or rubbing noses. Others would interpret someone's reluctance to make physical contact as insulting. It is a good idea to learn about the style of greeting which is more familiar to your client. If you feel awkward about adopting it, say something to the client. An international student once told me that she felt proscribed by her religion from shaking hands with males. The way she coped while living in the U.S. was to bow, bring her hands together, and say, "This is the greeting of my country." By discussing the difference in your styles, you will convey your respect and good will. This will probably make up for any inadvertent gaffes you made during the greeting.

3.4 Meetings Are Influenced by Non-Verbal Communication

Once you and your client have greeted each other, it is time to begin your discussion. What you should say, how you should say it and when you should say it are important considerations, but they are topics of later chapters. The focus in this chapter is on the non-verbal behavior that communicates your intentions to your client. Research in the field of communications suggests that nonverbal communication is responsible for more than half of the meaning of the total message that you send to your client (Cordell and Parker, 1993). These nonverbal cues are very influential. If a client perceives a discrepancy between what you are saying in words and what you are communicating with your non-verbal behavior, research shows that she is more likely to believe your non-verbal message!

You can create a positive environment for discussing technical issues by indicating non-verbally that you are attending to what the client has to say. Non-verbal communication includes the distance you stand or sit from your client, the extent to which you make eye contact, your facial expressions, the posture of your body and the gestures that you make. These dynamic behaviors all provide cues to the client that you are interested and listening. It is also important to be aware of cultural differences in how these non-verbal behaviors may be interpreted.

Most non-verbal communication is learned and interpreted sub-consciously. You may be completely unaware of your expectations in non-verbal communication until you encounter a discrepancy in your interaction with someone else. These discrepancies can cause an emotional reaction such as discomfort, hurt or even anger. You may make generalizations about someone mainly on the basis of their non-verbal behavior. Problems can arise because the norms for non-verbal communication vary among people from different cultural backgrounds. For the purposes of this book, I want to define the term "culture" to refer to the social influences that have shaped a person's expectations for communication. This definition of culture allows for distinctions not only between people from different countries or different ethnicities, but also between people from different regions of the same country, and between males and females within the same society. People from different cultures may expect and interpret non-verbal behavior differently. When you are interacting with someone from another culture, you both may misinterpret each other's non-verbal communication. Your technical conversation may become uncomfortable and unproductive even though nothing negative is intended. You need to be aware of non-verbal communication in order to avoid some of the pitfalls that can sabotage a discussion about technical topics.

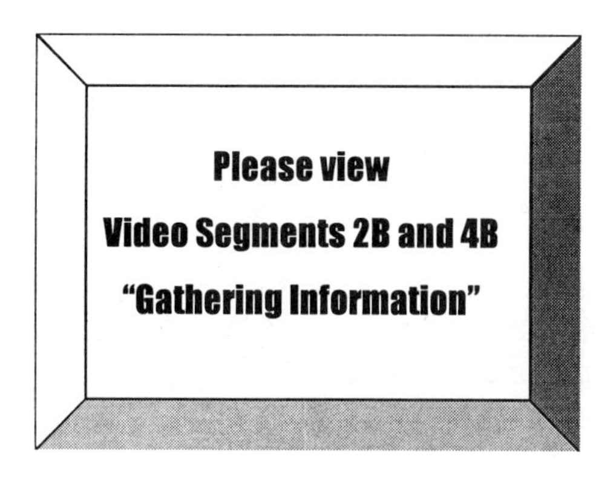

For a dynamic illustration of these non-verbal behaviors, take a look at video segments 2B and 4B. If you can, watch the segments now before continuing with this chapter. These segments depict Dr. Derr and Mr. Johnson in conversation together. Neither segment has sound. Without the sound, you can focus on the non-verbal behavior without being distracted by the technical content of the conversation. Segment 2B is an excerpt of the negative version of this meeting and segment 4B is an excerpt of the positive version. As you view each segment, think of words that would describe the impression that Dr. Derr is making on Mr. Johnson. For each of these descriptive words, specify how Dr. Derr's non-verbal behavior has created that impression for you. Then describe Mr. Johnson's reactions. Describe the non-verbal behaviors that conveyed Mr. Johnson's impression of the conversation. Finally, compare and contrast the positive and negative versions of this segment purely on the basis of non-verbal behavior. What prediction would you make about the outcome of each version of the meeting?

Non-verbal communication will influence the outcome of a meeting. You can convey your good intentions through non-verbal communication. You may also have your first clue that a technical discussion is not going well by observing your client's non-verbal communication. Because you and your client may have different expectations for non-verbal communication, you will need to recognize these reactions and adjust accordingly. Here is how you can use and interpret the major components of non-verbal communication:

Eye contact. Eye contact can be used to convey your interest and understanding to the client. In segment 2B, Dr. Derr made very little eye contact with Mr. Johnson. This made her appear bored and disinterested. The result of this behavior is that the conversation appeared to flow less comfortably. By contrast in segment 4B, Dr. Derr appeared more interested and understanding through the use of eye contact. The eye contact appeared to facilitate the conversation.

If your client is making more or less eye contact with you than you expect him to, you may begin to develop a negative reaction to him without really knowing why. As with other non-verbal behaviors, it is the departure from your expectation that leads to a negative reaction. For example, if your client is making more eye contact than you expect, you might begin to feel as if he is being overly aggressive or attempting to assert his dominance over you. You may even think he is making an unwelcome physical overture. On the other hand, if your client is making less eye contact than you expect, you might begin to think of him as passive or submissive. You could even begin to form an impression that he is not telling you the truth. These negative impressions will not help you and your client establish rapport with each other. Keep in mind that the use and interpretation of eye contact has a cultural context. In some cultures, eye contact is used to communicate equal status among the speakers and to encourage participation in the conversation. In others, eye contact can be used to indicate dominance. A person in a more submissive position would be disrespectful if he made a lot of eye contact. Just as you may be forming these negative impressions of your client, he may be making the same unwarranted assumptions about you!

Inter-personal distance. Inter-personal distance is the distance that two people will choose to stand or sit from each other. This distance is defined by each person's need for "personal space". You can think of personal space as the volume of space that surrounds and envelops each person. The boundary of this volume is defined both by cultural norms and by the nature of the relationship between two people. If someone sits or stands within somebody else's boundaries, we say that person's personal space has been invaded. As this term suggests, standing or sitting too close to somebody can make that person feel uncomfortable or even threatened. Sitting or standing too far from somebody can give the impression of being disinterested, inattentive or even angry.

The client-consultant relationship is most similar to a business relationship between people who, especially at the outset, are otherwise not acquainted with each other. For most cultures, the inter-personal distance in a business relationship is generally greater than it would be in a more personal relationship. However, across cultures, the norm for the inter-personal distance in a business relationship does vary. Problems can arise when your unconscious expectations are not met. Is the client sitting closer to you than you expected? If so, you might begin to feel uncomfortable. You could even feel threatened or that he is making an unwelcome physical overture. It is difficult to carry on a statistical discussion with these feelings of discomfort. Is the client sitting further away than you expected? Then you might begin to think he is not paying attention, or is unwilling to work with you. You may even wonder if he is angry about something. Monitor your reactions and consider that you or your client may simply be reacting to the amount of space between you. Permit your client to adjust his distance from you. However, consider your comfort also. If either one of you feels that your personal space is being invaded, your technical discussion will probably suffer. This is a circumstance where it is best for you to say something and request that you adjust your chairs to a position that is more comfortable for the person with the need for the greater inter-personal space. If this person is you, you can use other cues to indicate your interest and offset your need for greater inter-personal distance.

Body posture. An open posture, which has you leaning slightly towards the client, will help to convey your interest to the client. You saw good examples of this in segment 4B. A closed posture, with your arms and/or legs crossed and your body leaning away from the client, can all convey disinterest, as was evident in segment 2B. Keep in mind that in some cultures a very relaxed and open posture is acceptable, and in other cultures the same posture would convey disrespect.

Your client's non-verbal behavior can be an early warning system for problems. If you notice that your client has adopted a closed posture, you could take this as a signal that he may be uncomfortable about something. At that point you can look for other clues that could indicate discomfort also. There may be no problem at all: your client may simply prefer a more formal posture than you are accustomed to. The problem might simply be that your non-verbal behavior is contrary to his expectations: perhaps you are unintentionally invading his personal space, or you have adopted what to him is a disrespectful posture, or you are making too much eye contact for him to be comfortable with you. There may be a more serious problem. For example, there might be a difficulty with the statistical part of the project. Perhaps the negotiations about the consulting arrangement are causing a problem. Maybe he does not understand what you are saying and is embarrassed to admit it. Regardless of the cause of the problem, one of your first clues may be your client's non-verbal behavior. By staying alert to these cues, you can identify potential problems early and then address them before they grow too large.

Just as you benefit from observing your client's non-verbal behavior, you can also benefit by observing your own first impressions. When you are able to take note of any negative impressions without instantly believing them, you are in a better position to assess what is causing them. They may be caused simply by departures from your own expectations for non-verbal communication. These may be quite correctable so that you can continue to establish rapport and develop a good working relationship with your client.

3.5 Creating a Good First Impression Across Cultures

When you are planning to meet with someone whom you believe has a different background from yours, consider doing the following:

- Keep in mind that many first impressions are non-verbal, and that these impressions have a cultural context.

- Learn about the norms for your culture, especially for polite conversation between people who are not acquainted. What is the typical greeting? What is the norm for eye contact and body posture? What is the inter-personal distance for sitting and standing?

- Learn about the norms for these same non-verbal behaviors in the cultures of the clients with whom you will be working.

- Where possible, adapt your behavior to the client's norms. For example, you might adopt the style of greeting that is the norm in your client's culture and then follow this up with the greeting from your culture. If your client prefers to sit a little bit farther away from you than you would normally expect, permit her this extra distance.

- Become skeptical of any negative first impressions you develop about your client. Do you think she is not telling you everything because she is making less eye contact with you than you expect? Consider that perhaps she is simply showing you respect as a person of higher status. Do you think he is being pushy or aggressive? Perhaps instead he is accustomed to sitting closer to people than you are.

- Observe your client for any signs of discomfort. These will often manifest themselves non-verbally by a closed posture (client is turned away from you; arms and/or legs can be crossed) and averted eye contact. You will also probably be making little or no progress with your technical discussion.

- Discuss any signs of discomfort or any differences in cultural norms that you feel will interfere with your communication. Most clients will appreciate your consideration, and you can begin your technical discussion with a greater sense of rapport.

3.6 Interpreting Non-Verbal Communication in a Group

Non-verbal communication is very important in a group setting also. We will consider the small task-oriented group as our model of a group meeting in statistical consulting. A small group is especially well suited to make use of non-verbal communication, because each person can observe and be observed by each other person. This provides each member with instant feedback about his or her participation. Facial expressions, gestures, use of space, position and posture, will reflect each individual's attitudes, their reactions to others, and the overall progress of the meeting.

Non-verbal communication is especially important in small group meetings as a means to regulate the conversational exchanges among members. For example, if you are speaking in a meeting and the other members show by various non-verbal means that they are paying attention, are interested, or agree with what you are saying, you will probably be likely to continue to speak. Similarly, if you begin to speak and others show disinterest or disagreement non-verbally, you are more likely to limit your remarks. A small task-oriented group is most effective when everyone contributes their ideas while staying focused on the problems that need to be solved. This means nobody should waste the group's time by talking excessively about irrelevant issues, and also that everybody should feel free to contribute to the discussion. Non-verbal communication is one way that a small group has of channeling the conversation, and it can be used to the benefit or to the detriment of the objectives of the group.

The story in Example 3.1 illustrates how non-verbal communication can influence the discussion in small groups:

Example 3.1

Dr. Ann Seder† is a statistician who is working with a group of scientists and physicians on a three-year research study. The best way to recruit and group patients and analyze the data had been a matter of debate and uneasy compromise throughout the course of the study. The entire research group of ten people has met three times a year to review progress and make decisions. At these meetings, Ann and Dr. Klerk, a physician in the group, have frequently

† The names and details of examples appearing in this chapter, unless otherwise indicated, are fictitious.

> disagreed with each other about these issues. The meeting we will describe took place towards the end of the study as the final decisions about data analysis were to be made. The ten members of the group sat around a long, rectangular table. Ann sat along one side of the table with Dr. Klerk to her immediate right. When the discussion turned to the analysis of data, Ann began to give her recommendations. Every time Ann started to speak, Dr. Klerk grabbed Ann's arm, interrupted her, and proceeded to disagree with her. As the meeting progressed, Ann attempted unsuccessfully to elude Dr. Klerk's grasp by edging her chair down the table away from him. However, she was obliged to stop before intruding on the space of the person to her left. As she became more uncomfortable with the conversation, she noticed that others around the table looked uncomfortable also. Meanwhile, the discussion about the analysis had reached an impasse.

Some obvious messages were being sent by non-verbal communication in Example 3.1. Dr. Klerk disagreed with Ann and was using touch to stop her from speaking. Ann was uncomfortable with his use of touch and tried to remove herself from the problem. However, she did not want to sit uncomfortably close to the person on her left. Other members in the group appeared to react to the discomfort that they observed.

The story in Example 3.1 also shows how non-verbal communication can be used to diagnose the condition of a meeting. At the end of the example this group appears to be in trouble. They are distracted from their research agenda. There is clearly some disagreement among members of the group, but not all members are being permitted to express themselves freely. The discussion is at an impasse. If you had been an invisible observer and had entered the meeting room at this point, you might have seen this tension reflected in the posture, the position, the facial expressions, and the eye contact of the members of this meeting. If instead you had walked into a different meeting, where all of the members were openly engaged in a problem-solving discussion about the best way to analyze the research data, you could expect the non-verbal behavior of members of this group to reflect their engagement in the open discussion. Non-verbal communication will give you a good idea about whether or not any problems exist with the way a group is functioning. It may be difficult to decide what the problem is, or how to address it, but at least you can be alerted to the existence of tensions that might impede the problem-solving abilities of a group.

What happened to the research meeting with Ann and Dr. Klerk? This is actually the topic of a later chapter when we cover how to handle difficult situations. We will revisit Example 3.1 in Chapter 8 and examine different versions of the conclusion. However, if you don't mind straying from the topic of non-verbal communication for a moment, here is the first version of the conclusion to the story:

> Finally Ann could edge to the left no further. Dr. Klerk grabbed her arm and interrupted her one more time, saying "I am not comfortable with the way you plan to analyze the data." She took that as her opportunity to say, "I am not

comfortable with the way you are grabbing my arm." Dr. Klerk immediately apologized and said that in his country it would be an acceptable gesture. Ann accepted his apology and asked him to explain his objections to her plan. This appeared to defuse the tension in the room, and the entire group engaged once again in finding the best course of action for the analysis of data from this study.

The surface problem, a cultural interpretation of the use of touch, was a cue that there was a more substantial problem, the conflict between Ann and Dr. Klerk. In this version of the conclusion of the story, discussing the surface problem permitted the two of them to re-orient to the needs of the group to find a satisfactory way of analyzing the data from the research study.

3.7 Non-Verbal Communication in Settings Other than a Meeting

Statistical consulting occurs in a variety of settings. You may be able to meet face-to-face with your client or clients, or you might be conducting your discussion via a remove medium such as the telephone. Settings for statistical consulting in the past ten years have been made more diverse by the development of other forms of interactive communication, such as email, FAX, and video conferencing. Today's statistical consultant faces the challenge of communicating effectively in person, in writing, on the telephone, by email, FAX and video!

So far, the focus in this chapter has been on the face-to-face meeting. This is the setting that permits the broadest access to non-verbal communication. All other settings are deficient in some way. You can't shake hands on the telephone. You can't dress professionally for a FAX machine. You can't make eye contact on email. Since much of the way you communicate your intentions and diagnose potential problems in the conversation is done by non-verbal cues, it is important to consider the limitations and possible adaptations you may be able to make with each medium of communication.

The telephone. In many jobs, a statistician is connected to her clients mainly by telephone. On the phone, the tone of your voice and the timing of your speech are the main non-verbal ways you have to convey your intentions. This is why telephone courtesy is so important. It may be difficult to tell how your client is reacting to you. If you can, find some way of meeting your client face-to-face at least once, to facilitate getting to know each other. This is especially important if you have some complex issues to discuss.

Electronic mail (email). Email has become increasingly more important as a means to exchange and disseminate information rapidly and widely. It is very convenient to be able to send and receive electronic files and hold conversations at any time with people all over the world. You can read your email whenever it is convenient and respond to it according to its urgency. Many statisticians use email to augment a telephone conversation. They can send statistical reports, tables, and figures by email and then discuss them on the phone. It is becoming more common for workplaces to have a networked computer environment. Within this environment it is easy for all members of a project team, regardless of their location, to view the same information on their computer screens while they discuss a project by telephone. However, despite all this convenience brought about by email, it is best to keep in mind its limitations. Email is written communication, even though there are conventions that make use of keyboard symbols to convey emotion :). This means that there are no nonverbal cues available in this medium. Emotion is difficult to convey and easy to misinterpret. There are many cautionary tales about the person who fires off an angry email note and then accidentally sends it to everyone in the company. Complex negotiations are also difficult. Email is best used to communicate neutral and factual information!

Video conferencing. A videoconference provides a visual and auditory medium for non-verbal communication. However, you will probably experience some limitations in the way you can interact with others nonverbally in a videoconference, depending on the actual set-up. The number of cameras, monitors, and the speed with which they can transmit and switch images will all affect the visual resolution available to the conference. It may be difficult for you to send and receive more subtle visual cues that ordinarily you would use to indicate that you are listening, or to diagnose how your client is reacting. Gesturing or other rapid movements are also problematic, depending on the transmission speed of the cameras and transmission line.

Exercise 3.1

Consider the norms for non-verbal behavior in your culture, country or region. Apply these norms to the situation where you are meeting with a client for the first time. If you are an international visitor to this country, indicate any differences you have already experienced or expect to experience in these norms. Use your ideas to fill out the form shown on the next page. If you have the opportunity, discuss what you have written with someone whom you believe may have different norms from yours.

3.8 Suggestions for Group Discussion

1. You can show the video segments covered in this chapter in class. If possible, invite clients and practicing statisticians to view the video segments along with the class. Watching the video with the sound off comes is a dramatic way to demonstrate the importance of non-verbal communication. As a group you can identify non-verbal cues that are interpreted differently and those which have fairly uniform interpretation. Analyze and discuss any patterns you can find among these similarities and differences. This is an excellent way to learn about how clients react to their statistical colleagues. Viewing the video material provides neutral ground for this important discussion, since no one person in your discussion group is being critiqued directly. The tapes also present fairly exaggerated behavior, so it is possible for everyone to have a laugh at the actors' expense while still letting each other know what behavior they find acceptable and what they find unacceptable.

Form to be used for Exercise 3.1:

Your reference culture, county or region:	
Non-verbal behavior important in statistical consulting	**The norm for your culture; any comparisons you can make to norms in this country (if relevant)**
How late you can be for the first meeting without offending the client	
How late the client can be for the first meeting without offending you	
The way you greet the client	
The distance you and your client sit apart from each other at a table	
The amount of eye contact that you make with your client	
The amount of eye contact that your client makes with you	
How you indicate agreement or understanding with what your client has said	
What form (if any) of touching is acceptable	

2. Exercise 3.1 should enable you to explore the cultural differences and norms for non-verbal behavior among the participants in your class. You may be fortunate enough to have several cultures represented in your class. After people have had a chance to work on their own, form them into culturally diverse groups of two

or three. After 10-15 minutes, have a spokesperson from each group report their findings. Invite anyone to volunteer any experiences they have had in the area of missed cues or expectations with non-verbal behavior. If you do not have a very diverse group, invite some individuals to visit who would not object to discussing their expectations for non-verbal communication.

3. If you have videotapes of consulting meetings, view some of them with the sound off. Have the class interpret the body language. This is another opportunity to emphasize the differences in the interpretations given to nonverbal behavior by people from different cultures.

4. If students are not very familiar with interpreting body language in group meetings, you can refer them to Chapter 8 of Gulley and Leathers (1977). In this chapter on non-verbal communication, the authors present 10 photographs of group meetings that have been staged in order to represent different types of group dynamics.

3.9 Resources

Copeland Griggs Productions (1987), *Valuing Diversity Part III: Communicating Across Cultures,* San Francisco.

Gudykunst, W.G. and Kim, Y.Y. (1984), *Communicating with Strangers: An Approach to Intercultural Communication,* Menlo Park, CA: Addison-Wesley Publishing Company.

Gulley, H.E. and Leathers, D.G., (1977). *Communication and Group Process: Techniques for Improving the Quality of Small-Group Communication,* NY: Holt, Rinehart and Winston.

Hall, E.T. (1977), *Beyond Culture,* Garden City, NY: Anchor Press.

Hamilton, C. and Parker, C. (1993), *Communicating for Results: A Guide for Business and the Professions,* Belmont, CA: Wadsworth Publications.

Kim, Y.Y. (1986), *Interethnic Communication, Current Research,* Newbury Park, CA: Sage Publications.

Smith, J., Meyers, C.M., and Burkhalter, A.M. (1992), *Communicate: Strategies for International Teaching Assistants,* New Jersey: Regents / Prentice Hall.

Ting-Toomey, S. and Korzenny, F. (1991), *Cross-Cultural Interpersonal Communication,* Newbury Park, CA: Sage Publications.

4

MEETING

4.1 Introduction

Statistical consultants are involved in many meetings! These can be large or small, formal or informal, planned or spontaneous, in person or by remote transmission. You may have attended meetings that you felt were not very productive, even a waste of everyone's time. However, a meeting with your client or with the project team is your opportunity to provide input on the project. At a meeting, you can make sure that your work is aligned with the project's goals and with the work of others. A well-organized and well-conducted meeting should enable you to make better use of your efforts on the project. This chapter is about what you can do to make sure that these meetings are productive.

4.2 Learning Outcomes

- Identify the components of a meeting.
- Describe your own preferred style of communication.
- Discuss how you would adapt your "one-on-one" discussion for client who has a different preferred style of communication.
- Compare and contrast the structure of a "one-on-one" meeting with a team meeting.
- Describe the adaptations necessary for a remote meeting compared with an "in-person" meeting.

4.3 Identifying the Purpose of a Meeting

As a statistician, you probably will be attending meetings that are organized for a variety of different purposes. Giving out information, making plans, brainstorming, solving problems, making decisions, negotiating, gathering feedback, hearing opinions, and developing a consensus are all reasons for meetings that take place during the life of a project. The purpose or purposes of a meeting will shape who should attend and what format the meeting should take. Here are some examples of typical meetings that involve statisticians:

Example 4.1

Neil Snowball†, a self-employed statistical consultant, meets in person with Jack Frost, a representative of the state's internal revenue service (IRS). The purpose of the meeting is to discuss a forecasting problem that the agency would like the consultant to work on. Mr. Frost will describe the objectives of the problem, the data that are available and the deadlines for the project. Neil will obtain the information he needs in order to provide a preliminary response about the project and then develop an estimate and plan of work to be submitted later.

Example 4.2

Neil Snowball later telephones Jack Frost, the IRS representative. The purpose of the telephone call is to follow up on the proposed budget and statement of work that Neil has mailed to the agency. Neil also wants some additional information about the data. Jack wants to modify the statement about permission to use data from the project. They will decide during this phone call whether or not to finalize a contract for this work.

Example 4.3

An organizational meeting of the WEST study is held at the start of the fall semester. The purpose of this meeting is to organize an evaluation of the effectiveness of the WEST program. "WEST" stands for "Women in Engineering, Science and Technology." It is an academic program designed to enhance the retention of women undergraduates in the nontraditional majors of science, mathematics and engineering. Attending the meeting are: Gillian Lewis, the director of the WEST program, Barbara French, the statistician, Jean Reina and Ann Kupinski, graduate students in statistics, and Tawanda Ray, an undergraduate intern. Prior to the meeting, Dr. French has prepared and circulated a file by email containing a preliminary agenda, asking for contributions from the team members.

Example 4.4

James High, an applied statistician at a large telecommunication company, is invited to attend a meeting on designing a telecommunication product specifically targeted for customers of some local markets. James and his boss, Mr. Jackson, representing an applied statistics group and Ms. Andrews,

† The names and details of examples appearing in this chapter, unless otherwise indicated, are fictitious.

representing an survey research group are invited to attend a product development meeting with members of one of the company's marketing group. The participants represent three business units of the same company.

Example 4.5

The sponsor of a new drug schedules a videoconference with a review team from the US Food and Drug Administration. The FDA review team is responsible for reviewing the new drug application. Representing the pharmaceutical company at this meeting will be a regulatory specialist, a statistician, and two of its medical staff. Representing the FDA will be the primary medical reviewer, a statistician, a specialist in pharmacokinetics and the director of the review division. Prior to the meeting, the sponsor faxes a draft of the agenda to the FDA. The FDA responds to this draft with an official letter concerning the agenda. Both letters become part of the permanent record of this approval process.

Example 4.6

The FOOD study is a multi-center clinical trial designed to investigate the effects of vegetable intake on important predictors of health in humans. The study consists of a sequence of feeding studies that will be held during a period of five years. The FOOD study holds a general meeting each year. These meetings are held in the Washington DC area. The meetings are organized by the coordinating center of the trial and last for two days. There are several purposes for these annual meetings: (1) to inform key personnel about the progress of the feeding studies and about any changes or news that might affect the study; (2) to collect feedback from the four field centers about the progress of the trials at their centers; (3) to present recommendations from several committees for discussion and decision-making. During some of these meetings, the coordinating center statistician makes a presentation concerning some aspects of the design or analysis of the studies. Discussion and decision-making follow the presentation.

Example 4.7

The Protocol Sub-Committee of the FOOD study holds regular telephone conference calls. The purpose of these conference calls is to discuss and make recommendations to the Steering Committee about the protocol for each of several trials that make up the study. Members of this committee are: the director of the coordinating center, the head statistician, the principal investigators and one additional representative from each field center. The conference calls are scheduled for every two weeks during the developmental period of each protocol. Between calls, drafts of the developing protocol are sent by express mail to each participant. Just before the conference call, an agenda is sent by FAX to each participant.

Example 4.8

The statisticians from the FOOD study have set up an informal discussion by email. The discussion group includes the head statistician from the coordinating center, the agency statistician, and the statistician from each field center. The purpose of this email discussion is to share ideas and perspectives about statistical issues that are relevant to the trials. For each topic, each statistician contributes as a reply addressed to all the other members of the discussion group. The email note grows with each new reply added to the other contributions to that topic. The head statistician is responsible for making the final recommendation to the Steering Committee about statistical issues but he tests his ideas first with this group and attempts to develop a consensus. Several of the statistical issues arising during the study generate an active email discussion with widely differing opinions.

These examples represent a diverse set of meetings that you might expect to attend as a statistical consultant. They can be broadly classified into one-on-one meetings and group meetings. This distinction is important because the conversational dynamics differ between a meeting that just involve you and your client, and meetings of larger groups. In a one-on-one meeting, you can be more flexible about the discussion. You have more opportunity to get to know your client and to shape the conversation according to your mutual preferences. A larger group meeting will probably have a leader, and this person will be more constrained to move the discussion along in a way that benefits the entire group. The group meeting may also be more formal than a one-on-one meeting, with a written agenda, rules of procedure and written meeting notes. Both types of meetings share a common set of components that promote effectiveness: (1) an opportunity to build rapport; (2) an agenda that clarifies the purposes of the meeting; (3) a discussion that stays on track with everyone participating; (4) a period of closure; (5) a follow-up summary of decisions and action items. However, because of the differences in formality and dynamics, we will address these components for one-on-one meetings separately from group meetings.

4.4 One-on-One Meetings

A one-on-one meeting gives you a good opportunity to learn about and adapt to your client's preferred style of communication. People differ in the way they prefer to converse with others. A *communication style* is a set of preferences that determine how one person feels the most comfortable in establishing rapport, interacting and expressing ideas to another. For example, in a work-related conversation, one person may prefer to start talking immediately about the business at hand, while another will consider it important to have a fairly extended period of non-business conversation, or "small talk", first. Gender, personality, and culture all influence a person's preferred communication style. In her book *Talking from 9 to 5*, Tannen (1994) discusses the differences between males and females in their conversations at work. Both genders make use of small talk to establish rapport and increase comfort in the workplace. However, men are more likely to discuss sports and current events and women are more likely to discuss personal concerns. Differences in style, when not recognized, can lead to misunderstandings that can sabotage a discussion. For example, someone may regard the person who launches directly into business as impatient or unfriendly. The person who spends a long time with small talk may be viewed by another as wasting time or unwilling to talk business. This is not a good beginning to a statistical discussion! It is important to learn about your own preferences and to be able to recognize the preferences of others. Once you recognize your communication style as just one of several possible styles, you can develop some strategies for adapting to the preferences of others so that these differences do not interfere with your effectiveness.

There are four dimensions of communication style that are especially important in statistical consulting. Each dimension is a continuum of preferences between two opposing styles (Copeland Griggs Productions, 1987). *Phasing* refers to the conventions people have for when to talk about certain topics. In a business or statistical consulting conversation, these conventions about phasing affect the timing and duration of small talk and business conversation. *Sequencing* refers to the way people prefer to structure the topics within a conversation. People on one extreme of the continuum prefer to start with one topic and finish discussing it completely before proceeding to the next topic. They perceive a logical sequence of topics in the conversation and prefer to follow the order of this sequence from beginning to end. People on the other extreme prefer to discuss several topics at once, branching off on tangents and returning more than once to topics in a circular or spiraling fashion during the conversation. *Specificity* refers to the way people prefer to arrange general and specific information about a topic. Some people prefer to begin with generalities and move to specific information, and others are more comfortable talking about specific details first and then moving to generalities. Finally, *objectivity* refers to the way people use language to convey their ideas. Some people prefer to use precise language to convey their meanings very directly. Others prefer to be more indirect, relying more on context and inference to get their ideas across. Phasing and sequencing are the two dimensions that affect the structure of a meeting, and so we will cover them in this chapter. The remaining two dimensions will be covered in later chapters.

Establishing Rapport

Your small talk about the weather, the latest sports event, or how difficult it was to find a parking spot or a baby-sitter can help to establish common ground between you and the client. This is part of the rapport-building phase of a conversation. Kirk (1991) points out that clients are often uncertain about what to expect from a statistician, and may be nervous about their own lack of statistical expertise. A period of small talk can provide some general reassurance that this conversation may not be as painful as they anticipate!

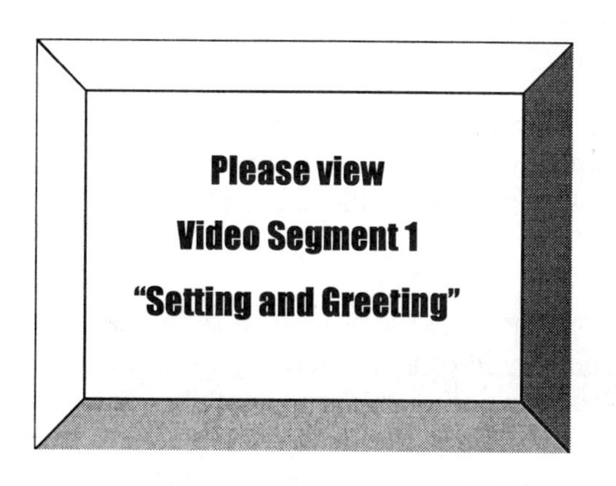

As you saw in Chapter 3, segment 1 of the video portrays a negative version of the greeting between Dr. Derr and Mr. Johnson. Take another look at this segment. If you are able to, review segment 1 now before proceeding with this chapter. Pay special attention to Dr. Derr's response to Mr. Johnson's attempt at small talk. You will notice that she did not respond in a very positive way to his comment about the encyclopedia. Instead, she got right down to business! We can guess that this version of Dr. Derr prefers giving a minimum amount of time to the rapport-building phase of a consulting conversation. She imposed this preference on her client by deflecting his remark about the encyclopedia and asking him what he was working on. What effect do you think this had on Mr. Johnson's feeling of comfort with the meeting?

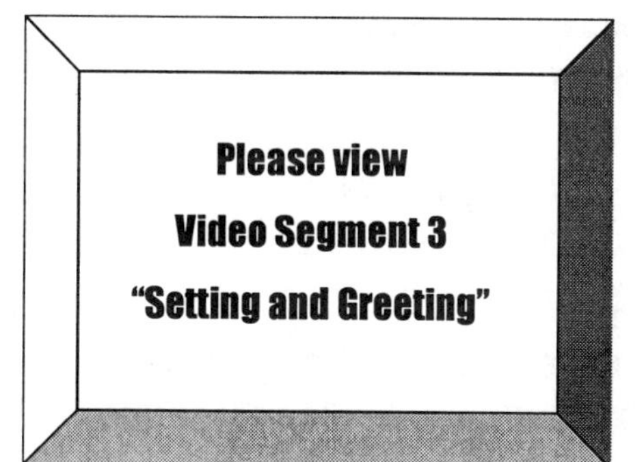

For a contrasting style, review the positive version of this greeting in video segment 3. In this segment, Dr. Derr makes a positive response to Mr. Johnson's comment about the encyclopedia, and then initiates some small talk herself. She only begins to talk about the project when Mr. Johnson appears to be more comfortable. This version of Dr. Derr is more attuned to the client's preferences for phasing.

You can become attuned to your client's preference for how much time he wants to give small talk and to business conversation. Consider in advance what you will do when you sense that you and your client have a difference in preference for phasing. If your client appears to prefer a longer initial period of small talk than you do, then extend this phase of the conversation if you possibly can. Statistical discussion is so challenging that it is important that both parties have developed a sense of rapport first. If you feel unable to extend the small talk phase, then find a polite way to move the conversation to more technical issues. On the other hand, if your client starts in on the business conversation while you would still like to get better acquainted, consider going along with this preference anyway. You may be able to interject some small talk later on. In your consulting career you are likely to work with people who have very different preferences for phasing than you. I always felt that I did not require a lot of small talk before getting down to business. However, this self-assessment was challenged when a client I had not met before strode into my office and began talking immediately about a project I had been assigned to review for her. I was scrambling mentally to figure out who she was and what project we were talking about, and developing a rather poor first impression of her. However, later on we discovered we were from the same hometown. Discovering how much common ground we actually had then helped me feel much more comfortable with this client.

Setting an Agenda

An agenda is a list of topics proposed for the meeting. The agenda serves to organize the discussion. When you and your client agree on an agenda, you then share responsibility for the progress and outcome of your discussion. This promotes a spirit of collaboration between you and your client.

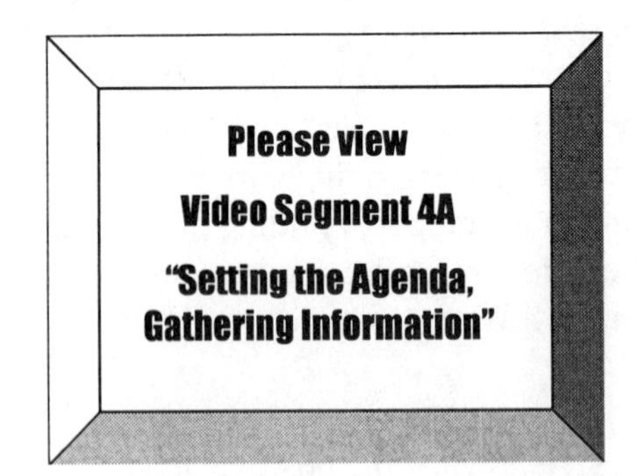

In a one-on-one meeting, the process of setting an agenda can be very informal. The first part of segment 4A shows Dr. Derr setting an agenda with Mr. Johnson at the beginning of their first meeting. If you are able to, view the first part of this segment now before continuing with this chapter. You can stop the video when the two of them have agreed on the agenda. What did you discover? Did they focus strictly on setting out the agenda before proceeding to discuss anything else? Definitely not! In fact this part of the conversation ranged fairly broadly. An excerpt of this conversation is shown in Exhibit 4.1.

Exhibit 4.1 ***Excerpt of agenda-setting conversation in video segment 4A***

1.	Dr. Derr	A meeting like this usually takes about an hour and I'm wondering if that fits in with your schedule.
2.	Mr. Johnson	Yes, that'll be fine … *[jokes about meetings]*
3.	Dr. Derr	I want to make sure that I understand enough about your study in order to make a good recommendation to you the next time that we meet. I wrote a few things on the board there, I want to make sure I find out about your study objectives and what your target population is. What else would you like to cover today?
4.	Mr. Johnson	Well, you know last week, … *[relates a story about an international student who visited the medical services] …*

		[both talk about this story]
5.	Mr. Johnson	Generally, I want to get a sense of what we are doing to help our client population.
6.	Dr. Derr	All right, so that's an overall objective, to find out how people feel in general about the medical services, and maybe to focus on the international students.
7.	Mr. Johnson	Yes, absolutely, and I also want to know, ...
		[more conversation about customer satisfaction surveys, committees, jokes about statisticians, limited statistical knowledge, trying to make this painless]
8.	Dr. Derr	So we'd like to talk about the sampling, getting the international students in there. Is there anything else we should cover?
9.	Mr. Johnson	Yes, I'm interested in what happens when people come in the door, how do they perceive our staff ... That's another thing, we don't have a lot of money.
		[jokes about the budget]
10.	Dr. Derr	Budget. So we should talk about that, too. Well, I'm just going to make a few notes on the board about that. *[writes]* That just helps me to stay on track. We don't have to follow that order in particular.

This may have seemed like a roundabout way to set an agenda! However, by the time they had finished this part of the conversation, they had established a time limit for the meeting, Dr. Derr had expressed her overall objective for the meeting, and they had each contributed two topics to the agenda. The tone of the conversation was friendly and sprinkled with humor. While setting the agenda, they were also continuing to get to know each other and build rapport.

Here are some other opening comments that you can use to invite an informal discussion about the agenda:

> *"What should we discuss today?"*
>
> *"I have about an hour available for this meeting. I know there is a lot we could cover. I'd like to make sure we get to the study design. What are the most important issues you'd like to cover?"*
>
> *"I see this as mainly a brainstorming session. You and I can throw out ideas about this study without being too critical of them just yet. We can write them down and then critique them later. How does that sound to you?"*
>
> *"I'd like to identify the response variables that you want to measure. I want to make sure I understand the priority you are giving to each of them. How much time to you have for this meeting?"*
>
> *"I have about 15 minutes to talk on the phone right now. What would you like to talk about? "*
>
> *"By the end of our discussion I want to make sure we're clear on what needs to be included in this report, and when you need to have the final copy. What issues would you like to cover?"*
>
> *"In this first meeting I like to make sure that I get a good understanding of your project and what you would like me to address. After we meet today, I would like to spend some time working on your request and then discuss my recommendations with you at a later time. How does that sound to you?"*
>
> *"I know we can't discuss the specific analysis until I sign the non-disclosure agreement with your company. But I think we can discuss the time line of the project and the limits for the overall budget. What would you like to talk about?"*

What these examples have in common are a respectful tone, a clear statement of what is important to the speaker and an invitation to the other person to contribute to the agenda.

Following the Agenda

There is no right or wrong way to organize a discussion. The way that you and your client follow an agenda will depend in part on how each of you prefers to sequence information. Unfortunately, problems can arise when one person prefers a more linear style of sequencing information and the other person prefers a more circular style. Each person can begin to form a negative impression of the other. "How controlling and unimaginative!" thinks the circular person about the linear person. "How illogical and time-wasting!" thinks the linear person about the circular person. These negative attributions can get in the way of a productive discussion. Fortunately, you can head off difficulties with this clash in styles. Once you understand your own preferences, you are in a better position to become attuned to the preferences of others. You can then adapt your conversation accordingly.

The fact that Dr. Derr wanted to list topics on a flip chart is a clue in segment 4A that she preferred a linear style of ordering information. The conversation that ensued immediately provided her with several clues that Mr. Johnson preferred a more circular style. He responded to her direct question about the agenda (comment 3 in Exhibit 4.1) with a story about an incident that took place at the medical services (comment 4). His subsequent comments were about a diversity of topics, although they all had to do with what he would like to find out from his survey. The way she adapted to his more circular style was to respond to each topic that he brought up, draw a generalization from it, and then return to setting the agenda. In this roundabout way, they were able to meet Dr. Derr's need to set an agenda while permitting Mr. Johnson the freedom to approach the subject tangentially from several directions at once. Dr. Derr's comments about the flip chart assured Mr. Johnson that she was not going to impose her linear style on him (comment 10). At the same time she asserted her need to keep the conversation on track (comment 10). In this way the conversation established a spirit of mutual adaptation, rather than a clash of styles.

If you also prefer an ordered sequence to your discussion but your client does not, you can try the approach depicted in segment 4A. You can work from an outline of a linear sequence of topics yourself without imposing it directly on your client. You can politely interrupt your client throughout your discussion and explain your need to develop your notes or to clarify your understanding. You can explain that this is how you are best at developing a good approach to the statistical issues. You can show your outline to your client periodically and show her how you are developing it from the conversation. You can then use the outline to identify areas that you haven't discussed yet. This approach will permit the client to converse in a style which is most comfortable to her while allowing you the structure that you need to do your best job.

Suppose instead that you prefer a more circular conversation but your client does not. If this is the case, you could encourage your client to keep track of the discussion from an outline. You could explain that you do your best problem solving if you are permitted to talk around the topics in a more circular fashion. You and your client could list some of the most important topics in a way that is visible to both of you, to make sure that you get to them in the conversation.

Coming to Closure

When the discussion ends, it should be clear to both you and your client what decisions have been made, what actions need to follow from the meeting, and what the circumstances will be for the next contact between you and the client. These elements are all part of closure. Although you may summarize and re-state decisions and action items throughout your conversation, it is very useful to revisit them at the end of your discussion. In this way you can make sure that you and your client have the same understanding of the main outcomes from your conversation.

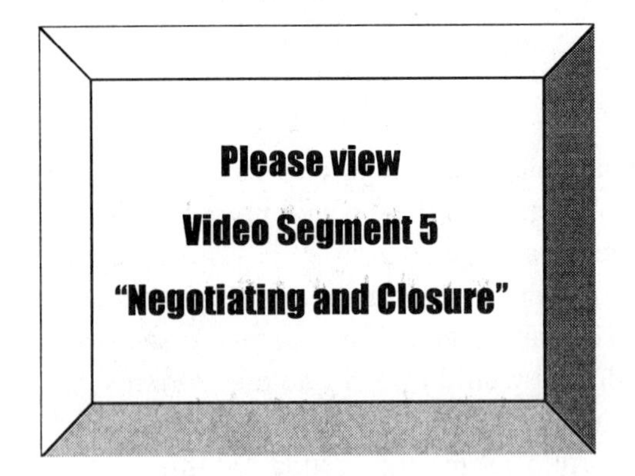

If you are able to view segment 5 now before continuing with this chapter, please do so. This segment is titled "Negotiating and Closure", and the elements of closure come at the end of the segment. When you view segment 5 for this chapter, focus your attention on the way that Dr. Derr and Mr. Johnson ended their meeting. An excerpt of Dr. Derr's comments at the end of the segment is in Exhibit 4.2:

Exhibit 4.2 ***Closure of the video meeting***

Excerpts of the close of segment 5, showing Dr. Derr's comments:

1. So, in terms of things that I need to do, I need to develop some estimates for you and consider advice about a sampling scheme, and how you might go about conducting this survey, particularly reaching the international students in a budget-friendly kind of way.

2. From your point of view I've asked you about specific figures for the populations, the subgroups of interest and then if it's feasible to do some kind of electronic random sample from the whole student body population.

3. So maybe you could email that to me later in the week and I'll try to get back to you as soon as I can.

4. Is there anything else that you think we should cover today?

5. Well, I think I got what I want, too.

In an informal, conversational way, Dr. Derr and Mr. Johnson summarized and agreed upon what each of them needed to do and by when. They then reflected on whether there were any more topics to cover, and concluded that they had had a satisfactory meeting.

Summarizing Your Discussion

A written summary of your discussion will help to reinforce and clarify the various decisions and action items that you and your client discussed during your meeting. This will take up more of your time, but think of it as an investment in your effective participation in a project. While writing up a summary of the meeting you may discover that some items are unclear, or that you and your client interpreted some decisions differently. It is best to find this out soon after the meeting than to let a misunderstanding grow and create discord. Most clients and consultants are busy with multiple projects. They may not accurately remember the discussion and decisions made at a meeting. And most projects, regardless of how carefully they are planned, encounter unexpected changes. When you keep a record of the discussions you have with your client, you can refer to that record to document when and why different decisions were made. This record may even serve to protect you in some circumstances.

Meeting notes do not need to be very formal. You can send your draft to your client for his or her input, and then make sure that you and your client both have a final version in your records. Exhibit 4.3 shows the email note that Dr. Derr sent to Mr. Johnson as a follow-up to the meeting that ended with segment 5:

Exhibit 4.3 ***A Meeting Summary Sent by Email***

Brian, thank you for coming to my office yesterday for our meeting. I think this is an interesting project and I look forward to working with you on it. I just wanted to clarify our "to-do" list that we put together during our meeting.

I had asked you to provide me with the size of the student population who would visit your clinic. It would be most helpful if I could have the size of the four groups we talked about: [graduate, international students], [undergraduate, international students], [graduate, domestic students], and [undergraduate, domestic students].

You had said you would telephone or email these numbers to me, and then from this I could prepare some possible sampling plans. You also said you would check with your committee to see whether they wanted our unit to help with the data processing and analysis.

Meanwhile I will develop an estimate of the cost of this work, based on the draft questionnaire you gave me. I will also look at the draft questionnaire and give you my reactions to it. Does this sound okay with you? I believe our target was to have this done by late next week. I'll call you next week to see how if we are ready to set up another meeting. Feel free to call or email me if you have any questions.

4.5 Meeting with a Team

Suppose that you have just been invited to attend a meeting with a project team. The invitation is a victory in itself! It means that the group has recognized the importance of involving a statistician in the discussion of their project. Examples 4.4 through 4.8 all describe team meetings. What should you expect to happen in this type of meeting? How can you best promote the good statistical practices you would like the group to adopt? Many statisticians have had very little experience in working with groups. If they have had a traditional, lecture-based education, their classroom training may not have prepared them well to meet and work in groups. Many statisticians are also somewhat introverted and prefer to work on their own. That is why the focus of this section of Chapter 4 is on what you can expect to happen in a group meeting, and how you can contribute most effectively.

When more than two people are involved in a discussion, there is usually somebody who takes the leadership role. The leader of a meeting has several responsibilities: (1) To organize and circulate the agenda; (2) To make sure that at least the high priority topics in the agenda are covered; (2) To make sure that everyone feels free to participate; (3) To make sure that the discussion stays on track; (4) To make sure discussions about controversial or sensitive issues are carried out with basic respect given to all. This is a challenging set of responsibilities! You may be a leader in some meetings and a participant in others. In a collaborative team, the leadership functions are often shared, either informally within the meeting or else by formally designating a different leader at different meetings. For these reasons it is important for you to learn how to be both an effective leader and an effective participant.

Setting an Agenda

It is easier to manage with an informal agenda when the meeting takes place between two people. When more than two people are involved, a more formal agenda is useful to organize the meeting. Several of the examples in this chapter include a description of a written agenda. When possible, it is a good idea to ask for input about the agenda from those who will be attending the meeting. In this way all participants are more likely to feel committed to the purposes of the meeting, and that their input and contributions are valued. Exhibit 4.4 shows the written agenda for the WEST meeting described in Example 4.3. This had been circulated by email to all participants in advance of the meeting for their input.

Exhibit 4.4 ***Written agenda for the WEST project***

WEST Program Evaluation
Women in Engineering, Science and Technology
Agenda for 10/3/96

Project Initiation Tasks
- Establish communication
- Establish conventions for project documentation

Project Task 1: Questionnaire
- Refine questionnaire
- Decide on schedule for evaluation

Project Task 2: Focus Group Discussions
- Define the purpose of these discussions
- What groups should be included

Project Task 3: Study of Retention
- Definition of retention
- Sources of information
- Study design

Establishing Rapport

Meetings often begin with a period of small talk. Food and drink may be provided. You may have even attended meetings where an "icebreaker" activity took place at the beginning of the meeting. The purpose of this phase of the meeting is to encourage participants to get to know each other. This will pay off later if the discussion becomes controversial or sensitive, or if some "give-and-take" type of negotiation is required. People who have developed some empathy for each other are more likely to be cooperative and less likely to let anger get the best of them. If you are organizing the meeting, make sure that you encourage some way of building rapport at the beginning of the meeting. If you are a participant and these activities are awkward for you, try to go along anyway. You do not have to disclose anything that makes you feel uncomfortable. If the icebreaker requires you to reveal your childhood nickname, and your nickname was "The Nerd", just make up a better one, like "The Champ". A good strategy, especially if you are shy, is to try to get to know some of the other participants one-on-one before the meeting. Stick next to them at the beginning and they will introduce you to the others.

Remember, to have been invited to attend this meeting is a very encouraging sign that the group values your input. Think about the benefits that may come your way once you need to discuss something difficult. You may get a better reception for your request for a much larger sample if the team has become better acquainted with you first.

Following the Agenda

The way a group moves through the topics in a meeting will depend a lot on the purpose of meeting. For example, your team may be "brainstorming" about a topic. This is a good way to collect creative ideas at the planning stages of a study as well as when a problem has arisen. The first step in brainstorming is to get as many ideas as possible recorded on a list without making any critical judgments about them. If you are leading a brainstorming session, you should make sure that you have created an open, non-judgmental environment for this part of the meeting. You may even do some preliminary exercises to make sure that the participants are feeling as creative as possible. Once the team has contributed as many ideas as they can, then you can have everyone examine the ideas more critically. If you are a participant, keep in mind that brainstorming is regarded as a creative, "right brain" type of activity. As a statistician, you may be more comfortable with logical, "left brain" types of activity. This session will give you an opportunity to do something different! Don't think too hard or critically about ideas during the brainstorming part of the meeting. Even if you think an idea is not perfect, contribute it to the list anyway. That is what the others in the meeting are doing, and it will show them that you are willing to expose your ideas to their scrutiny. Besides, seeing your "not very good" idea in the context of the ideas of others might just lead to the best solution to the problem. Once the session moves to the examination phase, you can then use your critical thinking skills to help the group sift through the whole set of ideas and identify a smaller set of feasible options.

In a group discussion, it is typical to discuss topics in the order that they appear on the agenda, although the leader may make allowances for team members who prefer a more circular style of sequencing. If you are the discussion leader, you should signal clearly where the discussion is relative to the agenda. You can indicate the start and the end of discussion on one topic and the transition to the next. At the point of transition you can summarize the discussion on that topic and indicate any decisions and "action items" for that topic. "Action items" are tasks that need to be done after the meeting finishes. Make sure that the person taking notes has recorded these decisions and action items. Finally, you can ask if there are any further contributions to that topic before you move on. If someone makes a comment that is out of sequence or not on the agenda at all, you have several options: (1) you can express your preference to proceed to the next topic but assure the person who brought it up that you will address his comment later; (2) you can ask the group whether they would like to proceed to the topic that has just been brought up; (3) you can say that the group won't be able to cover the topic in this meeting but you will attempt to address it later.

A good example of a team meeting and its subsequent follow-up begins with Example 4.4. In this example, two statisticians and a survey specialist from a large telecommunication company were meeting with representatives of the company's survey research group and three other business units of the company. The marketing managers, who had called the meeting, wanted to enlist the assistance of analytical specialists in the evaluation of the design of a telecommunication product. In the meeting, the marketing managers first gave some background and motivation for the development of the specific telecommunication service product for the local markets. They then listed many features of the service package to be tested for their desirability to the consumer. James High and his boss, Mr. Jackson were representing the applied statistics group and Ms. Andrews represented the survey research group. They asked many questions during this informal meeting, especially on the reasons behind each of the desired product features. Without getting into details of statistical design of experiment or marketing discrete choice modeling, the statisticians and survey specialists outlined in principle ways in which the desired task can be accomplished. The two analytical groups decided to work together to come back to the marketing managers with suggestions and a time and cost estimate of this evaluation project.

In the week following the meeting, James then phoned several marketing managers to clarify some specifics with this product. He became clearer on what was absolutely necessary in the different features' design and what was desired. Marketing managers at the same time had a better understanding on when they wanted to market this offer which implicitly imposed a hard deadline for the completion of the evaluation project. After some brainstorming by the members of these two analytical groups, James, along with Mr. Andrews, presented different avenues of evaluating the product and educated the marketing managers with the advantages and disadvantages of different approaches in terms of statistics, results interpretability, time and cost. There was some healthy disagreement among the methods suggested for this project. After several more conference calls and discussions, this working group jointly reached a decision on the approach to the analytical evaluation study.

You can see that the initial in-person meeting in Example 4.4 was just the starting point for the work that James needed to do for this project. His work involved further discussions by telephone with individual members of the team, and then a series of telephone conference calls for the group to come to a decision on the best approach to take for the evaluation study. The ability of the team to pursue actions and make decisions via telephone and conference calls was enhanced by their initial in-person meeting.

If the group is meeting via telephone conference call or videoconference, the discussion leader needs to help minimize the limitations of the medium. In general your actions will need to be more deliberate than they would be in an in-person meeting. Be sure to make introductions at the beginning of the session. Clearly signal the beginning and end of topics. At key points during the discussion, such as when a decision is being made, or at the transition between topics, take a poll of the participants and ask them for their input. If you think there might be some doubt about who just made a comment, identify that person. If you are a participant, you will also need to be more deliberate about your participation. At a videoconference, speak naturally but avoid distracting movements, such as shuffling papers, making nervous gestures or tapping objects near a microphone. Focus on the group discussion and avoid side conversations. On a telephone conference call, there will be no non-verbal cues available to indicate that you would like to speak, or that it is your turn to speak. Instead, you may have to start speaking as soon as someone has finished, without waiting for these non-verbal acknowledgments. This may seem somewhat rude compared to the way you normally speak in person, but it is the only way that the leader has to know that you'd like to make a comment. If more than one person starts to speak, the leader can then sort out who should speak next. It is a good practice to identify yourself at the beginning of your comment unless you are sure the other participants can identify you.

Coming to Closure

At the close of a meeting, the leader can summarize the highlights of the discussion, including key decisions and action items. This is a good time to make sure that participants are aware of what they need to do as a result of the meeting, and when and where the next meeting (if any) will take place. The leader can then ask for and provide feedback about how the meeting went, and collect suggestions about what might be done to improve the next one.

Summarizing Your Discussion

Once a meeting is finished, the person who was taking notes should write up a summary of a meeting to circulate to all of the participants. A written summary allows a project team to document the decisions and action items that affect the course of a project. These meeting notes are most useful if they contain the following elements:

- The date of the meeting
- The names of the participants
- A brief summary of each topic
- Decisions that were made during the meeting
- Action items: Things that require follow-up after the meeting. Be specific about who should do what, and by when
- Circumstances of the next discussion. You may know the time and date, or this may be conditional on some of the action items or other decisions.

An example of formal minutes of a meeting is shown in Exhibit 4.4. These are the meeting notes of the WEST project team whose agenda was given in Exhibit 4.3. The meeting notes were actually written into the file that contained the agenda outline. These minutes were circulated by email to the project team and then retained in the project notebook and electronic directory.

Exhibit 4.5 *Meeting Notes*

WEST Program Evaluation
Women in Engineering, Science and Technology
Meeting Notes for 10/3/96

Present at meeting: G. Lewis, B. French, T. Ray, J. Reina, J. Young

Project Initiation Tasks

- Establish communication

Tawanda has created a directory with the contact information for all project participants that is nearly complete.

ACTION: Gillian will give us directory information about her new staff assistant. When complete, Tawanda will test our email links by circulating the directory file to all project participants.

GOAL: By our next meeting 10/17/96 everyone will be able to send and receive documents and spreadsheets. Anyone who needs help can contact Tawanda for training and troubleshooting.

- Establish conventions for project documentation

Tawanda has started a project notebook with standard headings, and organized ancillary information. The group agreed to keep meeting notes and circulate them by email to all participants.

ACTION: Tawanda will document conventions for files, reports and computer directories, and circulate these to all participants before the next meeting.

Project Task 1: Questionnaire

- Refine questionnaire

Gillian is generally satisfied with the program evaluation form, and would be interested in having some comparable information for 1995, 1996, and 1997. However, we should consider special questions that could pertain to certain WEST cohorts.

ACTION: The entire WEST team will look at the program evaluation form from the above perspective and bring in suggestions to the next meeting. We will also look at evaluation material from similar programs from other schools.

... [more notes not shown here]

Project Task 3: Study of Retention

- Definition of retention

The group discussed possible definitions of "retention" as it pertained to this study. We need to develop a working definition of the changes in major that would qualify as a change out of a "science, mathematics and engineering" (SEM) major. These are not always well defined by the academic departments within the college, as we could think of a number of counter examples to each rule.

ACTION: Gillian will contact her colleagues in similar programs and propose a working definition of "change out of an SEM major" that we can apply to our data sources.

- Sources of information

We need to find out what data from student records is available and accessible to this project.

ACTIONS: Gillian and Barbara will find out what information is available from their respective departments. Jean will find out more about the college's data warehouse.

- Study design

We are considering a study design that enables us to identify both WEST students and similar (matched?) students and track their academic progress through several semesters. We will pay special attention to the event of changing out of an SEM major. We want a study design that will enable us to compare the retention in WEST students to the retention of similar students who were not in the WEST program.

ACTION: Barbara and Ann will look at possible study designs and evaluate their relative feasibility.

DATE OF NEXT MEETING: 11/2/96 3:30 p.m.

Exercise 4.1

The items below are dimensions that can be used to classify a person's preferred communication style. Each dimension is represented as a line connecting two extreme expressions of that dimension. The line represents a continuum of styles between the two extremes. Make a mark on the line that you feel represents your preferred position between the two extremes.

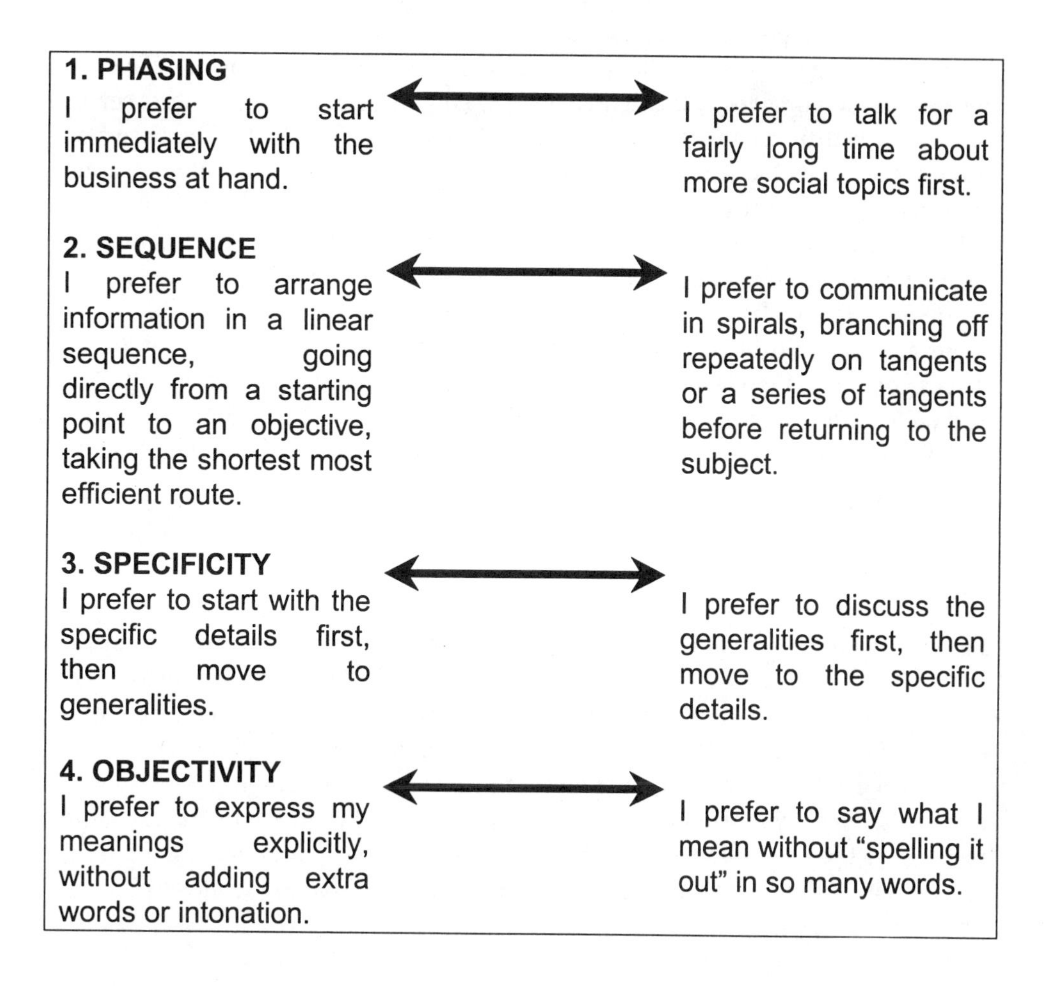

1. PHASING I prefer to start immediately with the business at hand.	⟷	I prefer to talk for a fairly long time about more social topics first.
2. SEQUENCE I prefer to arrange information in a linear sequence, going directly from a starting point to an objective, taking the shortest most efficient route.	⟷	I prefer to communicate in spirals, branching off repeatedly on tangents or a series of tangents before returning to the subject.
3. SPECIFICITY I prefer to start with the specific details first, then move to generalities.	⟷	I prefer to discuss the generalities first, then move to the specific details.
4. OBJECTIVITY I prefer to express my meanings explicitly, without adding extra words or intonation.	⟷	I prefer to say what I mean without "spelling it out" in so many words.

Exercise 4.2

Find someone in your class who marked their form differently for the dimensions of phasing and sequencing than you did. Discuss these differences. Come up with ideas about how you might recognize and adapt to these differences in a one-on-one conversation.

Exercise 4.3

Find an opportunity to view one or more recordings of actual meetings between a consultant and a client. If your unit does not have any such recordings, you may have to make one yourself. Be sure to get permission from the parties involved! Review the recording(s) and analyze the structure of the discussion: (1) Identify the rapport-building phase of the meeting. Note that this might take place at different points throughout the meeting. (2) How would you characterize the client's and the consultant's preference for phasing? What cues did you use to come to your conclusion? (3) Did the client and consultant set an agenda? Identify the agenda-setting parts of the discussion. Note that these also might occur throughout the conversation. (4) Characterize the consultant's and the client's preferences for phasing and sequencing. What cues did you use to come to your conclusion? (5) How did the consultant and client proceed through their agenda? Can you identify any adaptations that the consultant made to accommodate the client's preferred style for sequencing? (6) Identify the action items and other closure activities. Again, these can take place throughout the discussion. Was there a summary of these items at the end of the meeting? (7) Is there anything you might have done differently, had you been the consultant and had a chance to "replay" this meeting, in building rapport, following the agenda, and coming to closure? Please discuss.

Exercise 4.4

Find an opportunity to participate in a group meeting. As in Exercise 4.3, take note of the structure and processes of the meeting: (1) Describe the rapport-building phase; (2) Describe the agenda and the process of discussion through the agenda. How did the leader handle departures from the agenda? (3) Identify the action items and other closure activities. Was there a written summary? (4) If you had been the team leader and had a chance to "replay" this meeting, is there anything you might have done differently in replaying the meeting, in building rapport, following the agenda, and coming to closure? Please discuss.

4.6 Suggestions for Group Discussion

1. Here is a way to turn Exercise 4.1 into an in-class exercise: Assemble the class into small groups of 2 to 4, encouraging them to include members that represent diversity among communication styles. Each group can address one of the style variables from Exercise 4.1. Ask them to find ways of identifying someone's preferred style from a consulting conversation. They can brainstorm about how a consultant could adapt to a

client with a very different communication style. A spokesperson from each group can report the results of the discussion.

2. For an excellent depiction of people with a variety of communication styles, have the class view the videotape *Valuing Diversity. Part III: Communicating Across Cultures* by Copeland Griggs Productions (1987). Although the conversations in this videotape take place in general business settings, ask the class to consider the ways in which they would be similar to discussions between a statistician and a client. You can also have them identify any dissimilarity.

3. Provide an opportunity for the class to review recordings of actual consulting sessions. This will enable them to work through Exercise 4.3. If you do not have a library of such tapes, you could record consulting meetings that occur within your unit. Make sure you get the necessary permissions! You could also contact other consulting units to see if you could obtain tapes from them.

4. If possible, arrange to record members of the class during their meetings with clients. You can have them review portions of their tapes on their own, and then discuss portions of the tapes with you.

5. The best way for students to learn about the dynamics of team meetings is to get them involved in projects that have a team structure. When a student attends a team meeting, ask him or her both to participate and to pay attention to the structure and dynamics of the meeting. They can then report their observations to the class. With different students involved in different project teams, the class as a whole should learn about a wide range of types of meetings.

4.7 Resources

If you would like to learn more about teams and meetings, these resources can get you started. You can also find more references and training opportunities from your local human resources unit, the yellow pages of your telephone book, and the Internet.

Anderson, K. (1992), *To Meet or Not to Meet,* Shawnee Mission, KS: National Press Publications.

Brilhart, J.K. and Galanes, G.J. (1989), *Effective Group Discussion*, 6th ed, Dubuque, IA: Wm. C. Brown publishers.

Copeland Griggs Productions (1987), *Valuing Diversity Part III: Communicating Across Cultures,* San Francisco, CA

Kirk, R.E. (1991), "Statistical consulting in a university: Dealing with people and other challenges," *The American Statistician,* 45, 28-34.

Scholtes, P.R. (1988), *The Team Handbook,* Madison, WI: Joiner Associates.

Tannen, D. (1994), *Talking from 9 to 5,* NY: William Morrow and Company, Inc.

Yeatts, D.E. and Hyten, C. (1998), *High-Performing Self-Managed Work Teams: A Comparison of Theory to Practice,* Thousand Oaks, CA: Sage Publications.

5

ASKING GOOD QUESTIONS

5.1 Introduction

Statistical problem-solving begins with questions. What are the objectives of your client's study? What type of study is this? At what stage is the investigation? What limits and constraints govern the study? As you can see in Exhibit 5.1, the process typically begins with the client describing the problem that he would like you to address. As you talk with the client, you will probably begin to develop a more abstract formulation of the problem. This abstraction will then lead you to a statistical representation: a linear model, a probability formula, or some other statistical form. Your conversation provides the basis for this statistical model.

Exhibit 5.1 ***The Process of Statistical Problem-Solving***

Statistical model
Statistical solution
Client's problem
Client's solution

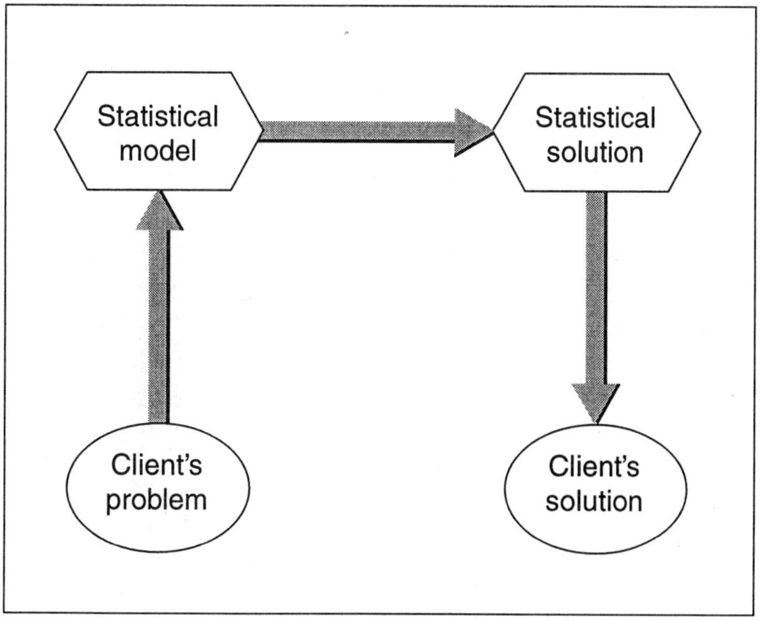

Once you have developed a statistical model of the client's problem, you can begin the work of developing a statistical solution, based on this model. You then have the job of translating your recommendation back to the client in language that he can understand, accept and apply to his problem. Although Exhibit 5.1 shows one loop of this process, it can actually be repeated many times as you and your client work together towards a solution.

Gathering information from your client about the technical nature of his investigation is one of the key tasks in statistical consulting. You need to know enough about your client's field of application to understand what the key statistical issues are in that field. You need to know what questions to ask and how to ask them in order to get accurate and complete information. Otherwise, you may leave the discussion with incorrect assumptions and a limited understanding of some crucial aspects of the problem. Mastering the art of gathering information requires time, knowledge and experience. This chapter is longer than the others, because learning what questions you need to ask and how to ask them well is relatively complex. Because of its length, this chapter is divided into three parts:

Part One: Avoid Errors of the Third Kind!
Part Two Identify What You Need to Find Out
Part Three: Develop an Effective Strategy for Gathering Information

5.2 Learning Outcomes

- Identify three main types of research investigation.
- List the important statistical issues associated with each type of investigation at the design and analysis stages .
- Describe how to learn about the client's field of application.
- Identify a strategy for asking questions that leads to accurate and complete information.
- Describe how to recognize and adapt to differences between you and your client in the way you prefer to address general and specific information.

5.3 Part One: Avoid Errors of the Third Kind!

What should a statistical consultant find out about a study? This is a very big question! Students and practicing statisticians alike are very concerned about what they should learn from a client in order to provide the best statistical input. There is no better place to start than with the paper published by A. W. Kimball in 1957 titled "Errors of the third kind in statistical consulting". Kimball may have been the first to coin the phrase "Type III error" which is now familiar to statisticians. This error can be defined as "Providing the right answer to the wrong problem".

One of the examples that Kimball provides in his article is quoted in Example 5.1:

Example 5.1

An engineer was engaged in particle size determinations in connection with corrosion studies. He wanted to estimate the particle size distribution, which he was willing to assume normal, but his method prevented him from observing particle sizes below a certain diameter. He knew very little about statistics but he had heard that there were ways of estimating distributions when samples are restricted. There was no statistician in his own group to whom he could turn for help, but there was one nearby who, although very busy, might give him a reference.

So he visited the statistician and presented him with the following sample of particle sizes: 25.6, 7.1, 5.1, 4.2, 3.7, 3.0, 2.6, 2.0, 1.8, 1.6, 1.5, 1.4, 1.3, 1.2, 1.1, 1.0, 0.9, 0.8, 0.7 -- and pointed out that his method would not allow him to determine particle sizes less than 0.7. Assuming the distribution normal, he wanted to know how he could estimate its mean and variance. The statistician was indeed quite busy and not inclined to spend much time on a problem he knew very little about and which did not originate in his group. On the other hand he did not want to cause any ill feelings by refusing to give any help at all. An easy way out was simply to hand the engineer one of his many reprints on truncated normal distributions (after all the engineer had asked for a reference), and this he did. Both participants in this short conference went away happy, the engineer because he thought he had an answer to his problem and the statistician because he disposed of an uninteresting problem in short order. But, as any reader who carefully inspected the "sample" of particle sizes already knows, an error of the third kind was committed. It might easily have gone unnoticed indefinitely, as do many others, but fortunately this error was caught.

The engineer returned to his desk armed confidently with the newly acquired reprint and began to apply the method with the help of his 1935 model calculator. He had not gotten very far along before he found that one of the statistics he computed was far outside the range of a key table given in the reprint to facilitate solution of the equations. After checking for and finding no arithmetical inaccuracies, he reluctantly returned to the statistician who inwardly was not too happy to see the engineer back. This conference lasted longer than the first, and with great chagrin the statistician finally realized what a stupid blunder had been made.

Among the methods used in particle size distribution is one known as the sedimentation method. Briefly, it consists of the preparation of a liquid suspension of the material to be analyzed and the measurement of the decrease in concentration of particles at or above a particular level in the suspension as sedimentation proceeds. Under suitable conditions, Stoke's law can be used to compute the percentage of particles in the suspension having diameters greater than *d*, say, where the value of *d* is determined by the time elapsed after sedimentation starts. Thus the random variables are the percentages, and *d* is a fixed or independent variate. It was this technique that the engineer had used. The appropriate method of estimation is, of course, probit analysis or one of its counterparts, and the "truncation" is not a problem except insofar as it increases the errors of estimate.

If the statistician had been familiar with particle size methods, or even if he had carefully scrutinized the "sample" that was presented to him, the error could never have occurred. It might be argued that both parties to this near-fiasco were the victims of circumstance and not really responsible, but if we are honest we must admit that the statistician has a duty to be more careful in avoiding this kind of error than perhaps any other. If he commits an error of the third kind, he is no less at fault than the physician who inadvertently administers arsenic instead of aspirin.[1]

You probably can identify several problems with the consulting story offered in Example 5.1. The statistician was very busy and did not feel much sense of commitment, at least at first, to this client's question. He did not possess enough knowledge about this client's field to provide a good recommendation. He apparently found the client's problem uninteresting. He chose not to take the time to discuss the problem thoroughly with the client or to look carefully at the data. All of these circumstances contributed to an inaccurate statistical representation of the problem. This caused the Type III error.

Clients also recognize the importance of obtaining an accurate statistical representation of their research problems. As one client put it: "When … you get into the solving of the problem, it's very important that the assumptions that are made are very clearly stated. Some of them may not be consistent with or even acceptable to the person based on the way in which the data have been collected. So it's really important to know what the assumptions were."

Exercise 5.1

Contact an experienced statistician and ask him to relate an occasion when he committed a "Type III error". Find out what he felt caused the error and what he would do the next time to prevent it.

---

[1] From "Errors of the Third Kind in Statistical Consulting" by A.W. Kimball, pp 135-137. Copyright © 1957, American Statistical Association. Reprinted with permission.

5.4 Part Two: Identify What You Need to Find Out

What do you need to find out from a client so that your recommendations are accurate and useful? There are so many different fields of application and so many different areas of statistics that this is a big question to answer! However, C. Chatfield has provided some broad guidelines in his book *Problem Solving: A Statistician's Guide* (1995). These guidelines are based on typical types of investigation and stages that characterize many investigations.

5.4.1 What type of investigation is this?

This is a good place to start. Chatfield (1995) identifies three typical types of investigation:

(1) Designed experiments. A designed experiment usually involves some manipulation of a system by the investigator. He selects or recruits subjects, or, more generally, "experimental units," into the study, and then allocates them at random to groups. The experiment then consists of different treatments or conditions applied to these groups. After the study starts, the investigator will measure one or more responses of the experimental units. The selection of individual experimental units is not usually done as a random sample from a target population. Instead, units are usually recruited or selected from a conveniently available source. An example of a designed experiment is given in Example 5.2:

Example 5.2 ***The Chocolate Bar Study: A designed experiment***

The purpose of the Chocolate Bar Study[2] was to determine whether a single chocolate bar per day consumed as part of a healthy diet produced a clinically important increase in plasma cholesterol. Forty-two healthy young men were fed a "Step 1" diet (a healthy diet recommended by the American Heart Association) for 21 days. After that they were assigned at random to two groups of 21 subjects each. One group received the "Chocolate Bar Diet", which was the Step 1 diet with a chocolate bar added to it. The other group received the "Pretzel Diet", which was the Step 1 diet with an amount of pretzels which were calorically equivalent to the chocolate bar. The two groups consumed their test diets for 27 days. Following the 27-day test diet period, both groups returned to the Step 1 diet for a washout period of 21 days. Then they crossed over to the other test diet for a further 27 days. Therefore each subject got both test diets. Half of them consumed the Chocolate Bar Diet first, followed by the Pretzel Diet, and the other half consumed the Pretzel Diet first, followed by the Chocolate Bar Diet. The order in which a subject consumed the test diets was assigned at random. At the beginning and end of each diet period, blood samples were taken from the subjects to determine their levels of plasma

[2] See Kris-Etherton, et al., 1994.

cholesterol. Measurements of interest were: Plasma total cholesterol, LDL-cholesterol and HDL cholesterol.

(2) ***Sample surveys***. A sample survey usually involves the specification of a target population from which a random sample is taken. From there, characteristics or responses are measured from the sample. If human subjects are involved, the responses may be measured from a questionnaire. The statistical inference in a survey is directed to the target population. The story line for the videotape is taken from a sample survey. This survey is summarized in Example 5.3.

Example 5.3 ***University Medical Services Survey: A sample survey***

The survey that forms the story line of the video is part of a quality initiative of the University Medical Services (UMS) of a large university in the northeastern United States†. The client, Mr. Brian Johnson, represents a committee charged by the UMS to develop a survey of students who are eligible to use the university's medical services. The key goals are to find out: (1) How important are several features of the medical services provided to the students; (2) How satisfied are the students with each of these features. The committee also wants to find out some demographic and usage information from the respondents.

Mr. Johnson and the committee he represents wants to make sure they draw a representative sample from the target population of all students who could potentially make use of the UMS. In addition, they are especially interested in obtaining the opinions of international students at the university. There are several reasons for this that come from word of mouth and have never been looked into, for example: (1) International students may have different expectations of medical students than "domestic" students; (2) Because they are far from home they may have fewer options for medical services outside the university; (3) Because of language and cultural differences, they may not always understand how to make use of the UMS; (4) They may have experienced different treatment by the UMS staff.

As you will learn in the video, Dr. Derr recommends a random sample of the target population and a random sub-sample of the international students. Both samples were drawn by computer from the university's database. A questionnaire was mailed out to these randomly-generated samples.

(3) ***Observational studies.*** An observational study involves neither sampling nor randomization. Instead, all of the available data is assembled and used to develop a statistical model for that system. The model can be used to illustrate relationships among variables and develop forecasts for future behavior of the system. It is not unusual to select individuals, or "experimental units", for inclusion into an observational study on the basis of certain

† The names (except for "Dr. Derr") and details of the video story line are fictitious.

characteristics. This selection is usually motivated by interest in studying possible causal relationships between variables. An example of an observational study is given in Example 5.4.

Example 5.4 ***Women in Engineering, Science and Technology (WEST) Program Evaluation: An observational study***

The purpose of the Women in Engineering, Science and Technology (WEST) program was to improve the retention of women undergraduates in non-traditional science, engineering and mathematics (SEM) majors. Data on retention indicates that women tend to drop out of SEM majors early in their undergraduate careers. The WEST program placed first year undergraduate women in laboratories and other research units around campus where they could participate in research. Previous research on retention that indicated that women might benefit from this additional socialization and participation in a research team outside the classroom.

To evaluate whether or not the WEST program was effective in improving retention, an observational study was carried out. The WEST investigators wanted to compare the retention in SEM majors between WEST interns and students who were not WEST interns but were otherwise similar in several key respects. They did this by obtaining permission to access the college's data warehouse, which maintained current and historical information on all students. For each of the 96 WEST interns they obtained two "virtual twins," one male and one female, who were similar by the following matching criteria: (1) age; (2) year and semester of entry into the college; (3) academic area of their intended major; (4) race; (5) SAT scores; and (6) self-report of the amount of certainty they felt about their choice of major.

Once the male and female "virtual twins" were established for each WEST intern, the academic progress of all of the students in the study was monitored for as many semesters as possible. The events recorded were: (1) dropping out of college; (2) changing to another SEM major; (3) changing out of an SEM major.

In the course of your experience, you will encounter studies that do not fit neatly as a designed experiment, a sample survey, or an observational study. This is not surprising! Some well-designed investigations simply have the characteristics of more than one type of study. Others are flawed.

5.4.2 At what stage is this investigation?

Once you identify the characteristics of the study you are working with, you can begin to identify what else you need to find out about it. Another good question to organize your discussion at this point is "At what stage is this investigation?" Chatfield (1995) identified seven stages as typical of most investigations. In his book, he discussed the statistical problem-solving tasks that take place at each stage:

Stage 1: The investigator is formulating the problem. At this stage, the statistician can participate by asking good questions and helping to clarify the objectives of the study. She can begin to formulate statistical representations of the study, and present the investigator with options, costs and benefits of different strategies for meeting the objectives.

Stage 2: The data are being collected. The statistician can provide input on the best method for collecting data. She may present several options along with their costs and benefits, so that the investigator may make an informed decision. She may also be involved in any randomization procedures that are required.

Stage 3: The data are being coded and entered into a computer format and their quality is assessed. The statistician may be involved in specifying procedures for data coding, management and quality assurance.

Stage 4: The data are being explored. The statistician may be involved in diagnosis and remediation of data errors or departures from model assumptions. She may be helping the investigator to develop an understanding of the descriptive results of his investigation.

Stage 5: Model building and inference procedures characterize this stage. The statistician may be making formal use of inferential methods required for testing hypotheses and constructing confidence intervals. She may be developing and testing the fit of several candidate models.

Stage 6: The results are being compared with any other information available that pertains to the study. At this point, the statistician or the investigator may decide that more data are required to meet the objectives of the current study.

Stage 7: The results are being interpreted and communicated to others. The statistician may be helping the investigator relate the statistical outcomes to the original study objectives. She may be involved in writing and presenting the results to other statisticians, to others in the investigator's field, and to the general public.

Finding out the stage of the investigation will help you to identify the tasks that you may be involved in. This will also help you to begin formulating what you need to find out about the study. It was my experience while at Penn State University's Statistical Consulting Center that most investigators made their initial contact with a statistician at one of two stages in their studies: About 25% of them contacted the consulting center for the first time at the planning stage of their studies, before any data had been collected. The remaining 75% first contacted the SCC at the analysis stage, after all of their data had been collected. Hopefully, the increasing emphasis on the statistician as a fully participating member of a project team will increase the percentage of projects that involve a statistician at the planning stages!

The statistical tasks at the planning and the analysis stages are fairly different. Despite the diversity of projects that involve statisticians, it is possible to develop some useful generalizations about these tasks. These will differ depending on whether the investigation involves a designed experiment, a sample survey or an observational study. For this reason, I have simplified Chatfield's seven stages into two main stages, planning and analysis. The special features of each type of study at the planning and analysis stages can provide you with a general idea of what you may need to find out when you are faced with a new consulting problem.

Planning Stage: If you are fortunate enough to be involved in a project at the planning stage, you can participate in defining and clarifying the objectives. This will bring into focus the priorities of the study. With the objectives and priorities in place, you can then identify the best methods of collecting data. You will need to find out what resources are available and what limitations constrain the study in order to develop a practical plan. You may also have an opportunity to develop a preliminary plan for analysis and reporting. You can sketch out the format for summary tables and figures. You can also describe the statistical inference that could come from their study. This will enable the team to make sure that their plan will actually produce the information that they need to address the objectives of their study.

If your project team is working on a designed study, then you may be developing the design. You may produce several alternative designs and identify the advantages and disadvantages of each for the investigators to make an informed decision. Part of this development will include calculating the sample size(s) needed for adequate statistical power. You can also work with the group to find the most practical ways to apply randomization and masking where it is needed. The Chocolate Bar study was a designed experiment, and I was involved with the project team at the planning stage. The study was going to be very costly, and so efficiency coupled with adequate statistical power were important issues. Fortunately, there was sufficient data from previous research to define a clinically important increase in a subject's total cholesterol. There was also sufficient data on variability for me to present the costs and benefits of a crossover study design compared with a parallel arm design. After the crossover design had been selected, I provided a randomization plan for the crossover design and a plan for the masked evaluation of all data from the study.

If the study you are planning is a survey, you may be helping the team define the target population and develop a sampling plan. Through your questions you will try to identify potential sources of bias that could threaten the validity of the statistical conclusions. You may calculate the sample size needed for the desired level of accuracy and statistical power. Your efforts at this stage to make sure that the data collection instruments are clearly worded and easy to follow will pay off with fewer errors and problems at the analysis stage. If you have the appropriate training, you may also help make sure that the questions and answer fields appropriately reflect the constructs that the team wishes to measure. As you will see in the unfolding story line of the video, Dr. Derr develops a sampling plan for the UMS survey. The details of the plan reflected the interests of the UMS committee to obtain a representative sample, stay within the budget, and make sure they have adequate representation of a minority subgroup within the sample.

If you are helping to plan an observational study, you may be identifying and evaluating the available sources of information. The databases may be very large and you may need to assess the feasibility of making linkages among them. The WEST Program Evaluation was an observational study. The statisticians obtained information about the data available with which to match WEST interns to non-WEST students. These data were available electronically in several locations both within the college's data warehouse and in a separate electronic repository outside the warehouse. They and the investigators discussed the variables that could and should be included. They also discussed criteria for matching subjects in order to develop acceptable ranges for each variable, a prioritized list of matching criteria, and a protocol by which to search and identify records for inclusion in the study. They worked with the investigators to obtain working definitions of the different outcomes of interest. These preliminary discussions enabled the WEST team to proceed to extract the data they needed relatively efficiently.

Analysis Stage: Let us suppose that the members of a project team have come to you for the first time after the data for their study had been collected or assembled, and they want you to participate in the analysis stage. Just as at the design stage, your first job should be to gain an understanding of the objectives of the study. Then you can evaluate how the data was collected, to see whether and how the objectives might be addressed. It is important for you to understand any limitations to the study that might affect the data analysis and subsequent statistical conclusions. Then you can develop a plan for the data analysis.

At the analysis stage of a designed experiment, you should find about how the treatments were allocated and applied to the experimental units. Were there any departures from the protocol during the course of the study? Are there any concerns about randomization? You should also gain an understanding of the response variables, how they were measured, how they relate to the objectives of the study and what their relative priorities are. Hypothesis testing and the construction of confidence intervals are usually very important in designed experiments, and you will probably need to specify what methods you plan to use. For the Chocolate Bar study, I examined the impact that some departures from protocol had had on the analysis. I also conducted a preliminary exploratory and graphical analysis of possible first period carryover effects on the data. This preliminary work led to the selection of the best linear model that represented the data and distribution of random error terms.

At the analysis stage of a survey, you will need to find out how the survey was implemented. You should identify any sources of bias and evaluate their impact on the study objectives. You should also investigate any other threats to the validity of the survey findings. In the UMS survey, a very low response rate was the main source of potential bias. Because of the low response rate, the study team decided against developing formal statistical estimates and confidence intervals that would have adjusted for the oversampling of the international students. Instead, Dr. Derr provided summary statistics in tables and figures to Mr. Johnson and his committee. The main challenge was to provide these statistics in a clear enough format for the committee to understand the results and to address their objectives.

At the analysis stage of an observational study, you may be involved in modeling the processes and relationships in the observational database. Linear regression models, loglinear models, survival models, neural networks, and linear structural equation models are just a few examples of the approaches you may take to represent the relationships in the database. Because of the observational nature of the data, many variables are likely to be confounded with each other to some extent. As Diane Lambert from Bell Laboratories pointed out:

> ... there is a new focus in industry on mining massive datasets that are not the result of careful experimentation or statistical monitoring or statistical sampling but are rather the results of reporting for administrative purposes or for corrective action or some other purpose that does not require careful thought about biases in data collection. The 'obvious' scenerio is that statisticians, continuing to work with engineers, will be asked to find the relationships in massive databases, and they will do so in ways that avoid reporting spurious relationships as certain truth.[3]

For the WEST Program Evaluation, the statisticians first produced graphical and tabular statistics of the retention data from the three groups (WEST interns, non-WEST males and non-WEST females). For the inferential analysis, the statistics team represented the matched groups in a logistic regression model. They emphasized that the interpretation of the results should include the reminder that the WEST interns had been a self-selected rather than a randomly-selected group.

If you would like to know more about any of these types of study and methods of analysis, there are some references listed at the end of this chapter to get you started.

5.4.3 <u>Learn about your client's field of application</u>.

One way to avoid making Type III errors is to become familiar with typical statistical issues that arise in your client's field of application. Kimball (1957) commented "The consulting statistician, particularly in the physical science and engineering fields, soon learns to question any 'corrections' applied by the experimenter before the data are presented for analysis." Every field of application will have its own statistical issues. The life sciences is a field in which I have a lot of experience. In this field it is very important for the statistician to identify all the sources of random error and statistical dependence in the client's study. This is because these factors contribute to the variation and covariation of the dependent variables. Understanding these sources and their relative magnitude will help to determine the most efficient study design and the most appropriate analysis. In a forum sponsored by the Committee on Applied and

---

[3] From "Discussion: Another View" by Diane Lambert, p. 202. Copyright © 1998, American Statistical Association.

Theoretical Statistics of the National Research Council, John Lehoczky made this point about communication across disciplines:

> In my view, true cross-disciplinary activities require that the statistician learn the language of the discipline and come to understand the fundamental problems of that discipline. This is in contrast to working with a subject-matter expert who translates the problem into the language of the statistician, often into a fairly precisely defined problem recognizable to (and perhaps solvable by) typically well-trained statisticians. I believe that the most important statistical skills in cross-disciplinary investigations involve structuring the questions to be asked and developing the methods of inquiry as opposed to being able to pull an especially appropriate statistical procedure off the shelf. The important activities must be carried out with an understanding of the discipline, its issues and methods themselves rather than in the language of statistics.[4]

How do you go about learning about a client's discipline? You may already have some background in another field. Many statistical consultants do have degrees in disciplines outside mathematics and statistics. However, if you are learning about a field for the first time, or you need to update your knowledge, a good source of information is the client herself. Find publications from work that is related to the current project. For more general information, find out if there are any key review articles or other resources that would be accessible to someone outside the client's field. You can start with articles that were written for the general public and proceed to more specialized references. You may also be able to find publications that directly address statistical issues in the client's discipline. You may be able to identify an appropriate course to take at a local college. If this is not an option for you, there may be short courses, workshops, or other distance learning opportunities that would enable you to get a better understanding of the client's field. A statistician who has experience consulting in the field you are interested in is another good resource. Discuss the statistical issues in that field with him and ask him to suggest other resources.

I hope the following story will help illustrate how to go about learning what you need to know about a client's field: When I joined a project team to work on a five-year grant in the field of biochemistry, I needed to gain a better understanding of the statistical and biochemical issues that would affect the study design and analysis. The title of the grant was "Cryopreservation of mouse sperm[†]." The principal investigator, Professor Roy Hammerstedt, from Penn State University's Department of Biochemistry, wanted to apply response surface methodology to identify optimal conditions for the long-term storage (by freezing) of sperm of genetically engineered strains of mouse. Professor Hammerstedt loaned me numerous references from the biochemistry literature that he thought would be helpful. Two of them pertained directly to statistical analysis in the field of the cryopreservation of sperm: "Issues in the statistical analysis

---

[4] From "Modernizing Statistics Ph.D. Programs" by J. Lechoczsky, p. 13. Copyright © 1995, American Statistical Association.

[†] Grant number HD-31761

of sperm motion data derived from computer-assisted systems" (Gladen *et al.* 1991), and "Variation of movement characteristics with washing and capacitation of spermatozoa. II. Multivariate statistical analysis and prediction of sperm penetrating ability" (Ginsburg *et al.* 1990). I also found an introductory level statistics book that was written for biochemists in this field titled *Flow Cytometry Data Analysis: Basic Concepts and Statistics* (Watson, 1992). I conferred with Professor Dennis Lin from Penn State University's Department of Management Science and Information Systems, who is an expert in response surface methodology (RSM). He generously provided me with some of his lecture notes. He also advised me that RSM was a technique that originated in engineering and related industry, and that an application of RSM to research in the life sciences would require more consideration to sources of random variation. He recommended that I obtain the reference by Box and Draper (1987) on *Empirical Model-Building and Response Surfaces*. This combination of reading and talking with experts enabled me to gain the knowledge I needed of biochemistry and statistics to participate in this project.

Clients also recognize how important it is for you to understand the important issues in their field. As one client put it: "Frequently it's been my experience from dealing with statisticians from a variety of different levels is that they have a wide diversity of knowledge as to what I deal with every day. If you sit there and don't say very much I may assume that you understand everything I say and I might make a quantum leap and start talking to you as if you were a colleague rather than a statistician. If you don't understand what I'm trying to do, and then we get to the assumptions of what I did, I don't know how you can understand that as well."

Exercise 5.2

A good resource for descriptions of studies is the book *A Handbook of Small Data Sets*, by Hand, et al. (1994). The *Handbook* includes 510 actual data sets, each of which is introduced with a brief description. If you can obtain this reference, select at least ten data sets and read their descriptions. For each one, identify what type of investigation you believe it to be and state why. Some descriptions do not contain sufficient information for you to classify them completely. For these, identify the information you would need to find out to complete the classification. Others may have some overlap or fit into more than one type of study. For these, describe what the overlap is and how this might have an impact on the analysis or the inference from the study.

Exercise 5.3

Choose a field of application that interests you, such as engineering, marketing, social sciences, forestry, and so on. With the help of others (for example, an investigator in that field, or an experienced statistical consultant in that field), find an example of a designed experiment, a sample survey, and an observational study in this field.

Exercise 5.4

Review files from previous consulting projects. For each project, identify the following: (1) the type investigation; (2) the stage(s) of the study which were addressed by the statistician; (3) the statistical representation(s) that were used. Put yourself in the place of the statistician and discuss how you would have avoided Type III errors.

Exercise 5.5

Choose a field of application that interests you. With the help of others, identify the following sources that you could use to learn more about important statistical issues in that field:

1. An article written for the general public.
2. A book or review article written at the introductory level.
3. An article about statistical topics in that field or a statistical text written for practitioners in that field.
4. A textbook written for statisticians that makes use of examples from this field.
5. A college-level course that you could take in that field.
6. A short course, workshop, or distance learning opportunity (such as an independent learning course or an Internet-based course).

Exercise 5.6

Interview a statistician who has experience in a certain field of application. Find out the following:

1. What special statistical issues are important in this field?
2. What should a statistical consultant find out from a client working in this field?
3. Are there any "pitfalls" that a statistician should watch out for?
4. Are there any good "rules of thumb" used in this field?
5. Are there any conventions in this field that affect the statistician's work?

5.5 Part Three: Develop an Effective Strategy for Gathering Information

Now that you have identified what you need to find out from your client about his project, it is time to consider how to go about asking your questions. Even though the information you require will vary from project to project, there will be a common need to obtain both a broad understanding of the major features of a study and detailed information about more specific issues. Therefore you will need to learn how to probe for general concepts and how to elicit specific details. How you introduce topics and coordinate your requests for general and specific information will depend in part on your and your client's preferred communication styles. As in previous chapters, you will see how to recognize and adapt to any differences in style that may otherwise interfere with your exchange of information.

5.5.1 Avoid poor communication strategies.

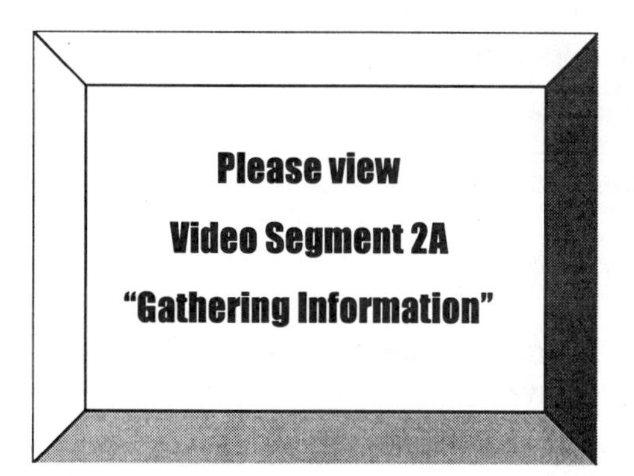

Asking good questions takes knowledge and practice. There are a number of "pitfalls" that can prevent good communication about statistics. If you can, view segment 2A of the video now before continuing with this chapter. What do you think about the questions that Dr. Derr asked Mr. Johnson? How would you describe his responses? Can you identify any Type III errors that Dr. Derr might commit as a result of this conversation? As you may have guessed, this is a negative portrayal of the way a statistician might ask a client about the facts of his study. It is an exaggerated portrayal of the way that many statisticians ask questions in an attempt to obtain technical information. The "pitfall" in this portrayal is the effect that a series of jargon-filled, closed questions can have on the client and the quality of information that the statistician obtains. A portion of this conversation between Dr. Derr and Mr. Johnson is shown in Dialog 1.

Segment 2A was not a very successful or comfortable information-gathering session! Dr. Derr did not take much time to interact with Mr. Johnson or to help him understand the statistical issues she was concerned with. She did not appear to acknowledge his remarks, but instead submitted him to a series of questions to which he needed to respond only with a brief answer. For example, leading into Dialog 1, Mr. Johnson had been elaborating the concern his committee had with international students who used the medical services. Dr. Derr's question #1 is an abrupt shift of topic. This question also contains the technical term "target population." Mr. Johnson did his best to supply his own definition (#2), since Dr. Derr did not provide any. She did not confirm his interpretation, but went on to ask another question about the composition of the target population (#3). In this question, she presented him with a forced choice. Her question implied that he could choose one of two options: that the target population was made up either entirely of undergraduate students, or of both graduate and undergraduate students. Mr. Johnson selected one of the choices (#4). Later on in their conversation, she asked a "yes/no" question about sampling that contained more technical language (#5). Mr. Johnson had no idea how to answer this (#6). Instead of clarifying her question, she lectured him about the importance of the target population (#7) to statistical inference. Later on in the discussion, Dr. Derr gave Mr. Johnson another forced choice concerning the form of the survey (#8): he could either select a telephone survey or a mail-in survey. When Mr. Johnson asked her advice, she delivered another lecture about the importance of this decision. These lectures did not appear to increase his understanding about the statistical issues of his project.

Dialog 1 ***Excerpts of the conversation in segment 2A***

1.	Dr. Derr	So what is your target population?
2.	Mr. Johnson	Uh, you mean the people we want to find out about?
3.	Dr. Derr	Well, you mentioned this forty thousand. Now is that undergraduate students on their own or is it graduates and undergraduates?
4.	Mr. Johnson	Well, it's graduates and undergraduates. …
		[more conversation]
5.	Dr. Derr	OK, so are you able to get a random probability sample of this target population?
6.	Mr. Johnson	A what?
7.	Dr. Derr	Well, that's going to be the basis of your inference, you've got this target population of forty thousand …
		[more conversation]
8.	Dr. Derr	Now is this going to be a mail-in survey or is it going to be a telephone survey?
9.	Mr. Johnson	Well which do you think would be more effective …
10.	Dr. Derr	Well, they are very different from each other …

A barrage of closed questions, forced choices, leading questions and unexplained jargon can seem like interrogation to a client. A sequence of poorly worded closed questions can have the effect of casting the client into a passive role in which he provides a few words of response to each question. Statisticians may do this in the mistaken belief that the more specific the question, the better the information they get. The problem with this approach is that you may miss the "big picture" -- context or factors that would make a big difference to your statistical response to the consulting problem. You may also be overwhelming your client with unexplained technical jargon and inappropriate forced choices. Your client may take the easy way out by taking his cue from the questions and providing the most desired response. He is less likely to contribute information and to feel responsible for the success of the exchange of information. You may not learn enough about the study to make an informed recommendation. In addition, the interrogation style of asking questions provides a poor model of communication to your client. When you need to discuss statistics with your client, he is less likely to respond by reflecting his understanding back to you. You would then have no idea if your client has understood the statistics. This is pointless!

Another illustration of a barrage of questions in a designed study is shown in Dialog 2.

Dialog 2

There is nothing wrong with questions that probe for specific details from the client. Statisticians do need to know about details in order to form a useful representation of the client's project. The problem comes when these questions are poorly worded or are used inappropriately. A leading question is one whose most desired answer is implied by the question itself. If a question is poorly worded, your client may be able to sense what the "most desired" short answer is, even if he doesn't really understand all of the technical terms in the question. For example, it may be clear to a client that the statistician in Dialog 2 would like to hear the answer "yes" to question 1, even when he is not sure what "at random" really implies about the process of assigning animals to groups. Or, the client may have used a process of assignment that he believes to be random but in fact has a serious flaw. A simple "yes" reply to question 1 will not inform you about potential flaws in randomization. A client may be inclined to give you the answer you want to hear, even when the circumstances of the study do not warrant the response. Most clients do not want to look stupid or lose face with you. They may not have understood the question, but they are pretty sure from the structure of the question what their answer should be. An incomplete forced choice question may imply to the client that only a few options are acceptable. In segment 2A, Dr. Derr presented two forced choice questions to Mr. Johnson, one about the composition of the target population and one about the format for the survey. In doing so, she may have missed important information such as non-degree students who may be part of the target population. She might also have closed down valuable options for conducting the survey, such as an in-person interview or an email survey.

When you ask a series of leading and incomplete forced choice questions, two things can happen. First, the information that you receive from this questioning technique may be inaccurate and incomplete. This can cause you to make a lot of assumptions about the statistical nature of the work that are not warranted. Second, this mode of questioning can move your client into a more passive role in which he provides very brief answers to your questions. Including undefined technical terms can further intimidate your client into adopting a passive role. In this circumstance you and your client no longer share mutual responsibility for the exchange of information. When you begin to discuss technical information with your client after subjecting him to this type of "interrogation", he may not be inclined to contribute or respond very positively to you.

In the past, statisticians may have had an unfortunate reputation for asking uncomfortable questions. In his article "The impertinent questioner: The scientist's guide to the statistician's mind," W. Lurie (1958) states:

> ... the statistician, if he is really to assist the scientist, must perform a necessary, but irritatingly annoying task: he must ask the scientist impertinent questions. Indeed, the questions, if bluntly asked, may appear to be not only impertinent but almost indecently prying – because they deal with the foundations of the scientist's thinking. By these questions, unsuspected weaknesses in the foundations may be brought to light, and the exposure of weaknesses in one's thinking is a rather unpleasant occurrence.

> The statistician will, then, if he is wise in the ways of human beings as well as learned in statistics, ask these questions diplomatically, or even not ask them as questions at all. He may well guide the discussion with the scientist in such a way that the answers to the questions will be forthcoming without the questions having been even explicitly asked.

Let us hope that public relations for statisticians have improved since Lurie wrote his article in 1958! However, it is well to keep in mind that the information that you must gather in order to do a good job is usually very important to your client also. There may be some sensitive issues concerning the statistical nature of the study. For this reason it is important to develop strategies that permit the exchange of information that is not only accurate but also comfortable.

5.5.2 Adopt more effective communication strategies.

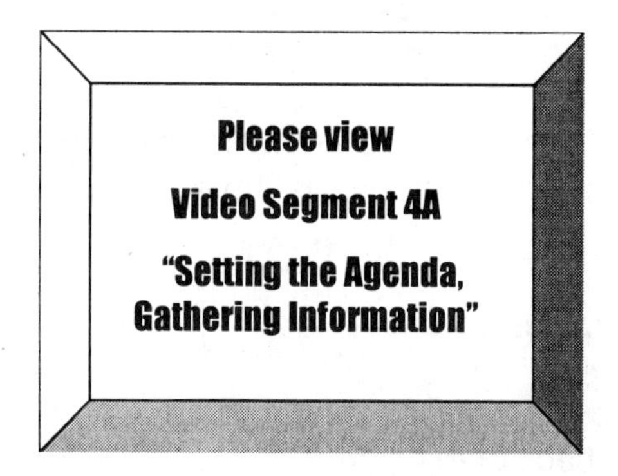

Now that you have looked at some examples of poor communication, let us consider how you can improve on these strategies. First, take a look at a better example of how a statistical consultant finds out information from her client. If you can, view segment 4A of the video now before continuing with this chapter. Focus on the exchange of information between Dr. Derr and Mr. Johnson. Consider how Dr. Derr improved her communication compared with segment 2A. Also, reflect on what cues you received in this segment about Mr. Johnson's preferred communication style in the dimension of specificity.

What you probably noticed right away is the slower pace of the conversation in the positive version. Dr. Derr permitted Mr. Johnson more entry into the discussion, and made more accommodation to the way he preferred to converse. Part of this conversation is transcribed in Dialog 3. In response to Dr. Derr's invitation to contribute to the agenda (#1), Mr. Johnson related an incident that occurred at the medical center (#2). Dr. Derr translated this story into an objective of Mr. Johnson's study, and asked for his feedback (#3). Mr. Johnson confirmed her interpretation and provided more information (#4), which Dr. Derr interpreted with the phrase "customer satisfaction survey," again asking for feedback (#5). In this way Dr. Derr identified the study objectives in terms that she could work with, making use of information from Mr. Johnson.

Dialog 3 ***Portions of the conversation transcribed from Video Segment 4A***

1.	Dr. Derr	I want to make sure that I understand about your study objectives and what your target population is. What else would you like to cover today?
2.	Mr. Johnson	Well, you know, last week one of our international students came in for our medical services and we had a really tough time trying to talk to them …
		[other conversation]
		… Generally, I want to get a sense of what we're doing to help our client population.
3.	Dr. Derr	Oh, all right, so that's an overall objective, is to find out how people feel about the medical services and maybe to focus on the international students in particular.
4.	Mr. Johnson	Yeah, absolutely … I also want to know if, when they come in and meet the receptionist, … is there a positive response …
5.	Dr. Derr	All right. So – sort of a customer satisfaction survey.
6.	Mr. Johnson	I like that [writes] customer satisfaction survey. It's something I can take back to my committee.
		[more conversation]
7.	Dr. Derr	So we'd like to talk about the sampling, getting the international students in there. Is there anything else we should cover?
9.	Mr. Johnson	Like I say, I'm kind of interested like I said in what happens when they come through the door, in how do they perceive our staff. You know we have about forty thousand students who are able to use the facilities at any given time …
10.	Dr. Derr	That's a big number.
11.	Mr. Johnson	It's a huge number. And two thousand of those are international students.

12.	Dr. Derr	OK, so, I would call that a target population, that forty thousand.
13.	Mr. Johnson	That's what that is. So that's forty thousand.
14.	Dr. Derr	That's the group of students that you would like to draw conclusions about.
		[other conversation].
15.	Mr. Johnson	... What we did come up with, I don't know if this is any help, but we did come up with this sample questionnaire
16.	Dr. Derr	It's very helpful, and you have to start somewhere, especially if you are working with a committee ...
		[other conversation]
		But it helps me understand what some of your issues are...
		[other conversation]
		So this is good, and what you'll find out is that you're going to be going through a lot of drafts of that questionnaire and pretesting it maybe in small groups ...

This brief conversation provided cues that Mr. Johnson preferred to speak about details first before moving to generalizations. His story about the incident with the medical center was his response to Dr. Derr's prompt about objectives, and then later in the conversation he mentioned some broader goals. Once this was evident from his conversation, Dr. Derr was able to paraphrase and summarize his remarks to make sure that she had the information she needed. With each paraphrase, she asked him for feedback to make sure she had understood correctly.

This paraphrasing style also helped her provide a definition of "target population" with Mr. Johnson's comment about the size of the student population (#9). The definition required several interchanges between the two (#9-#14) to make sure that it was appropriate and that Mr. Johnson had understood the concept. When Mr. Johnson showed Dr. Derr the sample questionnaire in this segment, she described the value it contributes to the project at the planning stage. This is in contrast to the rejection she gave it in segment 2A. She also provided Mr. Johnson with a brief overview of what the process of developing a questionnaire would be like. This is more useful information than was in the lectures she delivered to him in segment 2A. Overall, Dr. Derr took more time to interact with Mr. Johnson, to accept and work with his preferred communication style, and to give him more encouragement in segment 4A. The positive meeting probably also took more time overall to cover the same topics compared with the time required by the negative meeting. However, the investment in time to make sure both of them were clear with each other will lead to a better working relationship and fewer problems later on.

Now that you have looked at a more positive version of gathering information about a study, let us examine the specific strategies that can help you get the information you need about a study:

Use closed probes to get specific information. A closed probe is a question or comment that prompts for a brief response from the client. There is nothing intrinsically wrong with using closed probes, but you should use them only whenever you need to obtain specific details about a study. A common type of closed probe is a closed question. A closed question prompts for a "Yes" or "No" answer. This type of question will usually begin with the words "Can," "Did," "Do," "Are," or "Will." Items 1 through 5 are examples of this type of closed probe.

1. Can the animals interact with each other?
2. Did you interview every person on your list?
3. Do you have access to the 1950 census?
4. Are you including permanent residents in your survey?
5. Will all the soil samples come from the same tract of forest?

Closed questions are useful for obtaining very specific information, as long as they are clearly worded and free from jargon. Use closed questions when you want specific information, but keep in mind that the scope of the answer can be quite limited.

A forced choice question is another type of closed probe. Within the question there is a set of options from which to choose. The strong implication in this question is that the client should choose from among the options in the question. Items 6 and 7 are examples of forced choice questions. These questions can be used to elicit very specific information from the client. They should be used only be used when you are certain about the choices. Otherwise, you may be implying to the client that only certain options are acceptable. An incomplete forced choice question may cause the client to modify her answers to suit the statistician's definition.

6. Will you be looking at current records only or will you include information from the past?

7. Are you able to change the order in which different animals receive the treatments or must the order be the same for each animal?

Use open probes to get general information. Questions and probes differ in the response they elicit from a client. Some prompt for very specific information and others will promote a more general discussion. Open probes are questions or comments that prompt the client to discuss a topic more generally.

A request for information is a type of open probe. A request for information brings a general focus to the discussion while at the same time inviting the client to contribute her perspectives about the topic. A request for information often begins with phrases like "Tell me about ..." or "I'd like to know more about ..." Items 8 through 12 are examples of requests for information.

A request for information can also be used to close a topic and make a transition from one topic to the next. Items 13 and 14 are examples of this type of transition.

Open questions are broad, general questions that give the client a lot of freedom in responding. Open questions invite the client to provide a substantial response. They can also be phrased very neutrally and be kept free from technical jargon. Use an open question when you would like to have more general information than you can get with a closed question. Open questions usually begin with the words "Who", "What", "When", "How", and "Why". Items 16 through 19 are examples of open questions.

8. Tell me about this study. I'd like to hear about some of its objectives and the broad approach you plan to take.

9. I'd like to hear more about the responses you plan to measure.

10. Tell me a little about the background of this study.

11. I'd like to know more about the types of people you plan to target for the survey.

12. Tell me about how the animals will be maintained during the study.

13. Are there any other factors that might affect how people spend their money?

14. Is there anything more that we should cover concerning the layout of the soil samples before we go on to discuss the variables you plan to measure?

15. Who is likely to be excluded from the sample?

16. What factors are likely to affect the subject's response to treatment?

17. When do you plan to record information about food intake?

18. How did you assign the test diets to the animals?

19. Why is it necessary to run this group of simulations together?

As you have seen, an open probe prompts for general information, whereas a closed question prompts for a brief answer. A comparison between Dialog 4 and Dialog 5 below shows that the difference can be important:

Dialog 4

Consultant	Did you assign the animals to the groups at random?
Client	Yes. *(Thinks: What does "at random" mean?)*

Dialog 5

Consultant	How did you assign the animals to the two groups?
Client	I put the rats that were willing to exercise on the special treadmill in the "exercise" group, and the rats that avoided the treadmill in the "sedentary" group.

In Dialog 4, the consultant used a closed question that was leading and contained undefined jargon. The client responded with "Yes" as the most desired answer. In Dialog 5, the consultant used a neutrally phrased open question and gained a more accurate view of an important statistical issue.

> 20. Let me make sure I understand. This is a designed study, where you recruited subjects locally and then assigned them to the treatment or control at random.
>
> 21. So you will be targeting all students on this campus, and mailing out this survey to a random sample of them?
>
> 22. I want to make sure I understand what you have told me so far. You've gathered census and other economic data on the people in this state, and you want to forecast the future purchase of fishing licenses from this data.

Use concrete paraphrases to clarify your understanding. Communicating technical details is not easy. There may be several ways to interpret what a client has said about his project. It is important to make sure that you understand the technical details of the client's study so that you can make the most appropriate response. A concrete paraphrase is a restatement of factual information. It allows you to reflect to your client what your understanding is of what he has told you. Summarizing your understanding with a paraphrase will help you and your client make sure that you have correctly understood. If you have missed anything, a paraphrase will help the client to correct this misunderstanding. This will enable you to listen with greater accuracy. Items 20, 21 and 22 are examples of this type of concrete paraphrasing.

Using concrete paraphrases and open questions together is a very effective strategy for obtaining information that is balanced and complete. These open probes and paraphrases invite the client to participate fully in the conversation. This method also promotes a neutral and open tone to the discussion. This will help you to gain information about anything that is sensitive or problematical about the study. Dialog 6 is an example of how concrete paraphrases are used along with open questions to gain an accurate understanding of the client's study.

When you make use of open probes and concrete paraphrases to gather information, you also provide a model for the exchange of information in the opposite direction. When you present technical information to your client, the clearest indication that he has understood you comes from his ability to put your comments in his own words. By setting an example in the way you gather information, you will increase the chances that your client will also make use of concrete paraphrases. This will provide you with greater assurance that your client has understood your statistical comments.

Integrate open and closed probes. You will need to make use of both open and closed probes to get the general and specific information that you need. You can use open probes to introduce a topic, closed probes to get more specific information, and concrete paraphrases to summarize your understanding. Requests such as "I'd like to know more about the objectives of your survey" or "Tell me about how the animals were assigned to groups" are good ways to introduce

Dialog 6

The use of different probes is indicated by: Request for Information (RFI), Concrete Paraphrase (CP), Open Question (OQ).

Consultant	Tell me about how you designed your study. (RFI)
Client	I assigned 12 pigs to three diets: high, medium and low fat.
Consultant	So you had 36 pigs in all, with 12 assigned to each of the three diets (CP).
Client	No, there were only 12 pigs. Each pig had all of the diets.
Consultant	So each of the 12 pigs had all three diets (CP). What was the order in which the pigs received the diets? (OQ)
Client	In the first month I gave them the low fat diet, then the next month they got the medium fat diet, and in the third month they got the high fat diet.
Consultant	Let me make sure I understand. All 12 pigs were given the same order for the test diets, low - medium - high fat. (CP)
Client	*Yes, that was how we designed the study.*
Consultant	Why did you decide to use the same order for all of the pigs? (OQ)
Client	This kept things simpler for us, preparing only one diet at a time. I also felt that giving the high fat diet first might interfere with the animal's response to the other diets.
Consultant	I see. Your decision about the order was partly to keep the protocol simple and partly because of your concern about the high fat diet (CP).

a topic. You can also make use of open probes to signal a transition from one topic to the next. Questions such as "Before we move on to talking about the test diets, is there anything else we should cover about the housing conditions?" will signal your intentions to close one topic and move to the next one. At the same time, you give your client an opportunity to give you more information about the first topic if he feels it is important. Concrete paraphrases help you to correct and clarify your emerging understanding of a topic. You can summarize your understanding as a way of signaling the close of a topic. The paraphrase invites your client to make any final corrections necessary to the way you have understood him.

Exactly how you integrate open and closed probes into a conversation will depend on your preferences and your client's preferences in the dimensions of sequencing and specificity. As you learned in chapter 4, you may move through topics one at a time in a linear sequence, or cover several topics at once in a more circular style. The dimension of specificity also influences the way you and your client will organize a discussion. Some people may find the task of starting immediately with a discussion of broad purposes and concepts too abstract. They may prefer starting with more specific details and working towards the more general concepts. Others prefer discussing large concepts first and then moving to increasingly towards finer levels of detail. There is no right or wrong way to order general and specific information. Exhibit 5.2 shows how a discussion can make use of open and closed probes to move from general to specific, or to move from specific to general. However, if you and your client differ in this preference you may end up talking at cross-purposes. This is why it is important for you to pay attention to whether a client is discussing general or specific information. In video segment 4A, it was apparent that Dr. Derr preferred to bring up broad concepts while Mr. Johnson was more comfortable talking about details (as he did with his opening story about the international student). Recognizing this difference, Dr. Derr focused on paraphrasing Mr. Johnson's specific comments into generalizations, and then obtaining his feedback on the accuracy of her paraphrases.

If your client is concentrating on specific details, try to use concrete probes to remain specific. When you feel the need to understand more about the general context of the project, try making use of a concrete paraphrase that is a generalization of these details followed by an open question or a request for information. This will acknowledge what he has said to you and invite a more general discussion. On the other hand, if the client is speaking in generalities, use open probes to remain general. When you feel the need to focus on specific details, try making use of a concrete paraphrase of the generalizations followed by a closed probe. Again, this strategy acknowledges your client's comments while moving the focus of the conversation to more specific details. You may also consider varying your own approach as you move through your discussion. By moving from specific to general for some topics and general to specific for others, you may identify an approach that elicits the most positive response from your client. This will also enable you to increase your own range in the dimension of specificity.

Exhibit 5.2 ***Moving between general and specific***

Moving from general to specific

General

Request for information

Open question

Concrete paraphrase

Closed probe

Concrete paraphrase

Specific

Moving from specific to general

Specific

Closed probe

Concrete paraphrase

Request for information

Concrete paraphrase

Open question

General

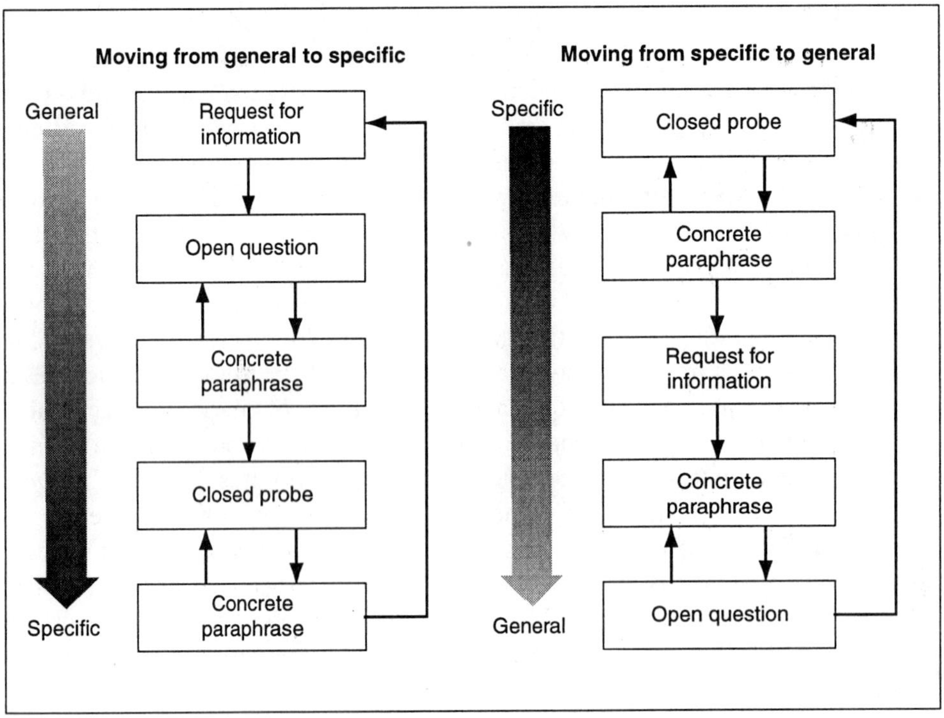

Dialog 7 represents a discussion that incorporates open and closed probes to move between general and specific information.

Dialog 7

The use of different probes is shown by: Request for Information (RFI), Concrete Paraphrase (CP), Open Question (OQ), Closed Question (CQ)

Consultant	I'd like to know how you selected people to participate in your survey *(RFI)*.
Client	I stopped them when they left the library and asked them to fill out the questionnaire.
Consultant	So you targeted people who had just used the library *(CP)*. When did you conduct these interviews? *(OQ)*
Client	I chose Monday and Friday at 10:00 a.m.

The use of different probes is shown by: Request for Information (RFI), Concrete Paraphrase (CP), Open Question (OQ), Closed Question (CQ)

Consultant	I see. You chose two weekdays at the same time in the morning *(CP)*. Why did you decide not to include other times of the day or weekends in the sample? *(OQ)*
Client	That was most convenient for me, and a lot of students do use the library at that time.
Consultant	If there were a lot of students coming out of the library around 10:00 a.m. *(CP)*, were you able to contact each one of them about the survey? *(CQ)*
Client	No, I just made sure that I got 25 students to fill it out each time.
Consultant	When you weren't able to contact them all *(CP)*, how did you choose which students to approach? *(OQ)*.
Client	I just approached students who looked like they would talk to me.
Consultant	Did some students refuse to fill out the questionnaire? *(CQ)*
Client	Yes, I'd say about half of them said they didn't have time to fill it out.
Consultant	Let me make sure I understand. At each sampling period, there were lots of students leaving the library and you selected students who looked approachable. You had to approach about 50 students each time to get 25 students who would agree to fill out the questionnaire *(CP)*.
Client	Yes, that describes how I got my sample.
Consultant	Is there anything else we should cover about your sampling method before we take a look at your questionnaire? *(RFI)*
Client	Well, I don't know if this is important, but some of my friends filled out questionnaires on both days.

Exercise 5.7

Return to Questions 1 through 7. Write an open version of each question.

Exercise 5.8

In Dialog 6, the consultant made a sequence of concrete paraphrases that reflected his emerging understanding of the study design. Identify the first statistical model that the consultant had in mind, and then the final correct one. How did the statistician avoid a Type III error through his use of concrete paraphrases? What design problems did he uncover through his use of open questions and neutral tone?

Exercise 5.9

View portions of any recordings you have available of your own consulting meetings or the meetings of other consultants. Assess the consultant's use of open and closed probes and the client's responses to these probes. Choose a part of the conversation and analyze the sequencing of topics and the movement of the discussion between general and specific information. Did you detect any differences in sequencing and specificity between the consultant and the client? If so, discuss how the consultant responded to these differences. Identify any occasions during the discussion when you felt that undefined jargon or other communication difficulties interfered with the consultant's understanding of the client's project. Did you feel that a Type III error might have occurred as a result of these difficulties? If so, state how you would have provided improved clarification.

Exercise 5.10

In his article "The questioning statistician," Finney (1982) has identified twenty-two questions that he believes a statistician should ask an investigator at the planning stage of a designed experiment. Critique these questions: What do you think would happen to your conversation if you were literally to ask these questions as written? Then develop a strategy for conducting an effective discussion with a client that would enable you to find out the information requested by these twenty-two questions.

Exercise 5.11

This exercise has ten brief hypothetical statements from clients. Use each statement to develop a strategy for gathering information. Identify what you would like to find out. Write out some open and closed probes and concrete paraphrases that you would use to gain and clarify this information. If it is helpful, include some client responses to your questions and comments. Feel free to improvise!

1. Each subject listens to all possible pairs of sounds in random order and states his preference.

2. We want to be able to forecast a drought from these daily measures of rainfall and groundwater runoff.

3. I will take a random sample from each box of mushrooms, evaluate their color, and then compute an average for each box.

4. We want to do a telephone survey to see what factors influence a customer's decision to purchase skim milk.

5. I have three brands of pesticide for three varieties of corn in three field sites.

7. I mailed out an opinion survey about the mandatory seat belt law in our state.

8. We are conducting a very expensive study on the effect of garlic on health, and we do not want to exclude subjects who are not willing to take the garlic capsules.

9. I want to survey people who like to go canoeing to find out how they select a state park for their recreation.

10. It was too much of a burden on the subject to exercise under all eight experimental conditions, so I assigned each subject to only four of the conditions.

5.6 Suggestions for Group Discussion

1. If you are working primarily with students who have not had much experience in consulting, including experienced statisticians would enhance the discussion. "Experienced" clients (clients who have worked with statisticians before) would also make a valuable contribution to this discussion. The process of translating information about a client's project into a statistical representation that is ultimately useful to the client involves many subtleties. A good resource for such a discussion is the article "Deconstructing statistical questions" by D. Hand (1994). Hand and the statisticians who served as discussants to the article all acknowledge the challenges inherent in applying a statistical representation to a client's problems. They provide a diverse set of examples from their collective experiences in statistical consulting.

2. Organize a class discussion around Type III errors. You can have the class interview experienced statisticians by telephone, in person, or email to complete Exercise 5.1. You can invite experienced statisticians to attend the class to discuss their experiences with Type III errors. If your class includes practicing statisticians, you can create some small group discussions in which participants share their experiences and then report to the class. This is a topic that could also be used to organize an Internet discussion.

3. You can encourage each student or small group or students to select different fields of application for Exercise 5.3 so that the class as a whole will be able to see how designed experiments, sample surveys and observational studies are used in a variety of fields.

4. You can help participants develop their mastery in identifying what they need to find out about a project. If you have past records of consulting projects, you can have the participants review these records. These resources provide the basis for Exercise 5.4.

5. You can organize the students in groups to locate the resources for learning about a client's field, as they are directed to do in Exercise 5.5. You can help them identify investigators and experienced statisticians who would be willing to help.

6. You can invite experienced statisticians to talk to your class on the questions in Exercise 5.6. If your class is mostly made up of experienced statisticians, you can organize them into discussion groups around the different fields of application represented by the class participants. Each group can identify a spokesperson to summarize the group's input to the questions in Exercise 5.6.

7. The questions in Exercises 5.5 and 5.6 can also be used to organize an Internet discussion. Invite experienced statisticians and clients to participate along with your class.

8. Exercise 5.9 makes the use of recorded consulting sessions. These can come from your library of taped consulting sessions or from the current discussions being conducted by the participants of your class. To get started you can show the videotape segments in class. Exercise 5.9 then requires a closer analysis of the dialog on the segments, which could be done individually or in small groups.

9. If at all possible, provide opportunities for the participants in your class to gain experience in consulting with actual clients. If you are able to arrange to record some of these meetings, you will have good resource material to help the participant assess his communication skills. You can have the participant view or listen to parts of his own tape and identify the elements of communication discussed in this chapter. This will give you an opportunity to provide some coaching about how these skills might be improved.

5.7 Resources

References on statistical problem-solving:

Chatfield, C. (1995), *Problem Solving: A Statistician's Guide*, London: Chapman and Hall.

Hand, D.J., Daly, F., Lunn, A.D., McConway, K. and Ostrowski, E. (1994), *A Handbook of Small Data Sets,* London: Chapman and Hall.

Joiner, B.L. (1982), "Practicing statistics, or, what they forgot to say in the classroom." In *Teaching of Statistics and Statistical Consulting* (eds. Rustagi, J.S. and Wolfe, D.A.), NY: Academic Press.

Kimball, A.W., (1957), Errors of the third kind in statistical consulting," *Journal of the American Statistical Association,* 52, 133-142.

Lurie, W. (1958), "The impertinent questioner: The scientist's guide to the statistician's mind," *American Scientist,* 46, 57-61.

Tweedie, R. (1998), "Consulting: real problems, real interactions, real outcomes," *Statistical Science,* 13, 1-29.

References about designed experiments:

Finney, D.J. (1982), "The questioning statistician," *Statistics in Medicine,* 1, 5-13.

Montgomery, D.C. (1976), *Design and Analysis of Experiments*, NY: John Wiley & Sons.

Meinert, C.L. (1986), *Clinical Trials: Design, Conduct, and Analysis,* NY: Oxford University Press.

Milliken, G.A. and Johnson, D.E. (1984), *Analysis of Messy Data, Vol. 1: Designed Experiments,* Belmont, CA: Lifetime Learning Publications.

Yandell, B.S. (1997), *Practical Data Analysis for Designed Experiments,* London: Chapman and Hall.

References about sample surveys:

Cochran, W.G. (1953), *Sampling Techniques*, NY: John Wiley.

Dillman, D.A. (1978), *Mail and Telephone Surveys: The Total Design Method,* NY: John Wiley & Sons.

Fink, A. (1995), *The Survey Handbook,* Thousand Oaks, CA: Sage Publications, Inc.

Hayes, B.E. (1992), *Measuring Customer Satisfaction: Development and Use of Questionnaires,* Milwaukee, WI: ASQC Quality Press.

Thompson, S.K. (1992), *Sampling,* NY: John Wiley & Sons.

References about observational studies:

Allison, P.D. (1984), *Event History Analysis: Regression for Longitudinal Event Data.* Newbury Park, CA: Sage Publications.

Agresti, A. (1984), *Analysis of Ordinal Categorical Data,* NY: John Wiley & Sons.

Bryk, A.S. and Randenbush, S.W. (1992), *Hierarchical Linear Models,* Newbury Park, CA: Sage Publications.

Cressie, N., (1991), *Statistics for Spatial Data*, NY: John Wiley & Sons.

Cheng, B. and Titterington, D.M. (1994), "Neural networks: A review from a statistical perspective," *Statistical Science,* 9, 2-54.

Draper, N.R. and Smith, H. (1981), *Applied Regression Analysis*, 2nd ed, NY: John Wiley & Sons.

Gilbert, R.O. (1987), *Statistical Methods for Environmental Pollution Monitoring*, NY: Van Nostrand Reinhold Co.

Kendall, M. and Ord, J.K. (1990), *Time Series*, 3rd ed, London: Edward Arnold.

McKinlay, S.M. (1975), "The design and analysis of the observational study – a review," *Journal of the American Statistical Association,* 70, 503-523.

Selvin, S. (1996), *Statistical Analysis of Epidemiologic Data,* Oxford: Oxford University Press.

Singer, J.D. and Willett, J.B. (1993), "It's about time: Using discrete-time survival analysis to study duration and timing of events," *Journal of Educational Statistics,* 18, 155-195.

Other references cited in this chapter:

Box, G.E.P. and Draper, N.R. (1987), *Empirical Model-Building and Response Surfaces,* NY: John Wiley & Sons.

Ginsburg, K.A., Sacco, A.G., Ager, J.W., and Moghissi, K.S. (1990), "Variation of movement characteristics with washing and capacitation of spermatozoa. II. Multivariate statistical analysis and prediction of sperm penetrating ability," *Fertility and Sterility,* 53, 704-708.

Gladen, B.C., Williams, J., and Chapin, R.E. (1990), "Issues in the statistical analysis of sperm motion data derived from computer-assisted systems," *Journal of Andrology,* 12, 89-97.

Hand, D.J. (1994), "Deconstructing statistical questions," *Journal of the Royal Statistical Society*, Series A, 157, 317-356.

Kris-Etherton, P.M., Derr, J.A, Mustad, V.A., Seligson, F.H., and Pearson, T.A. (1994), "Effects of a milk chocolate bar per day substituted for a high-carbohydrate snack in young men on an NCEP/AHA Step 1 Diet," *American Journal of Clinical Nutrition,* 60, 1036S-42S.

Lehoczky, J. (1995), "Modernizing statistics Ph.D. programs," *The American Statistician,* 49, 12-17.

Lambert, D. (1998), "Discussion: another view," *Technometrics,* 40, 201-203.

Watson, J.V. (1992), *Flow Cytometry Data Analysis: Basic Concepts and Statistics,* Cambridge: Cambridge University Press.

6

NEGOTIATING A SATISFACTORY EXCHANGE

6.1 Introduction

At the heart of statistical consulting is the reciprocal exchange that occurs between you and your client. You have agreed to provide something for your client, and in return your client has agreed to provide something for you. As you will learn in this chapter, the reciprocal exchanges in statistical consulting can often be fairly complex. Here are some examples that illustrate the variety of exchanges that can take place:

1) In exchange for an hourly fee, you agree to analyze data and produce a technical report by a given date.

2) In exchange for salary support, co-authorship, and the permission to present papers at a scientific meeting, you agree to bear responsibility for the design and analysis of a clinical trial.

3) In exchange for thanks and the prospect of future collaboration, you agree to provide a brief consultation over the telephone.

4) In exchange for expediting an analysis, your client agrees to include you in the planning sessions of the next study.

Negotiation is a process of communication. You use negotiation to come to an agreement with your client about what items will be exchanged. The goal of this chapter is to provide you with the communication skills you need to negotiate successfully with your clients. The emphasis is on achieving "win-win" outcomes in your negotiations with clients. As you will learn in this chapter, a "win-win" outcome is one in which all parties involved feel that their needs will be met. In this ideal outcome, everyone clearly understands the exchanges involved in the agreement. They all feel that the exchanges are fair. Mastering the communication skills in this chapter will help you to negotiate successfully even when the exchanges are complex and the client's preferred style of negotiation is markedly different from yours.

6.2 Learning outcomes

- Identify and characterize the key issues in statistical consulting that affect your negotiation with a client.
- Identify the tangible and intangible items involved in the exchange.
- Characterize the relative value of each item and the extent to which tangible and intangible items may be substituted for each other.
- Describe how to recognize and adapt to your client's preferred styles of negotiation and communication.

6.3 Put Negotiation on the Agenda

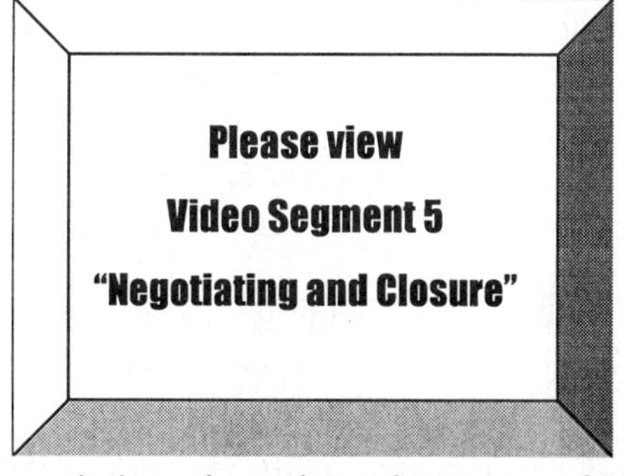

If possible, watch segment 5 of the video now before proceeding with this chapter. You will view an example of negotiation in statistical consulting. In segment 5, Dr. Derr initiates a discussion about costs, her role, and other issues that will affect her participation in the project. This discussion took place at the end of her first meeting with Mr. Johnson. By initiating a discussion of these issues at this early stage, she helped to establish negotiation as a topic on the agenda. She and Mr. Johnson can then revisit their negotiations throughout the course of a project in much the same way that they revisit their decisions about the statistical issues. This will help them adapt to any changes that may occur during the course of a project.

Experienced statisticians know that changes are the rule rather than the exception. Projects are dynamic. Discoveries along the way can alter your decisions about the best statistical practices for a project. In the same way, changes in the environment surrounding the project can also affect the set of agreements that you and the client feel is satisfactory. Make sure that negotiation is an agenda item for your ongoing discussions with your client. This will help you and your client continue to stay aligned on the most satisfactory arrangements for your participation in the project.

In segment 5 of the video, Dr. Derr and Mr. Johnson did not come to any firm agreements. They both needed to collect additional information. However, their first brief discussion served to identify issues, boundaries and possible exchanges that would define their future association.

Mr. Johnson was clearly concerned about cost. Dr. Derr acknowledged the funding limits and offered to provide several cost estimates based on different options. She probed to determine the acceptability of a lower-cost option that would take longer and involve a student intern. The student intern would need to have access to the data and permission to write up and present the results. In this discussion, Dr. Derr was attempting to identify the extent to which time and access to information could be substituted for money. Mr. Johnson was unable to give a definite answer although his initial reaction indicated concern about protecting the confidentiality of information about the respondents. Dr. Derr acknowledged this concern and offered to involve the client in the procedures that would be implemented to protect the confidentiality of the data. In this way Dr. Derr took her cues about the client's priorities and concerns directly from the conversation and identified possible options that could be developed to address these concerns.

6.4 Understand Your Client's Preferred Style of Negotiation

Although the entire first meeting leading up to segment 5 is not depicted on video, previous segments from this meeting showed how Dr. Derr prepared for this negotiation. You saw in segments 3 and 4A how she began by developing a good rapport with Mr. Johnson. The welcoming setting of the meeting, the respectful greeting, the positive non-verbal behavior and the open, collaborative style of questioning all served to create a comfortable environment for negotiation. These preliminaries can be very important to the success of your negotiations. In fact, for someone who prefers a "high context" style of negotiation, these preliminaries are essential. Raymond Cohen coined the term "high context" in his book *Negotiating Across Cultures* (1991). A person with a high context style of negotiation will draw many inferences from his surroundings, from nonverbal cues, and from hinted nuances of meaning. Cultivating a personal relationship before entering into a negotiation will be very important to him. He will be careful to avoid embarrassment, both to himself and to anyone else involved in the discussion. He may also try to avoid direct confrontation and be very reluctant to say a direct "no". A person with a high context style of negotiation will tend to let previously established conventions determine his position on many issues. He may not be inclined to modify a position if it is based on strongly held principles.

In contrast, a person with a low context style of negotiation will be likely to consider modifications of most issues. In fact, she may deliberately start out by stating a position that is more extreme than the one she actually feels. In doing so, she will expect that she and the other party will move back and forth between extremes and arrive at a compromise agreement that is somewhere in the middle of their two initial positions. This process is also known as "haggling". A person with a low context style of negotiation is likely to focus on the discussion at hand and not pay much attention to any previous history of negotiations or other previously established conventions. She will not be offended by contradictions or by confrontations. She will tend to value accuracy in language and prefer to speak very directly. A person with this style would not expect much preliminary rapport-building or other activities that help to establish a personal relationship. These two styles of negotiating are summarized in Exhibit 6.1.

Exhibit 6.1 ***Low and high context styles of negotiation***

A person with a low context style of negotiation:	A person with a high context style of negotiation:
1. Prefers to start negotiating immediately without establishing a personal relationship.	1. Prefers to establish a personal relationship entering into a negotiation.
2. Obtains meaning mainly from verbal discussion.	2. Draws many inferences from surroundings, nonverbal cues, and hinted nuances of meaning.
3. Is willing to "haggle" about most issues, including very important ones.	3. Will avoid "haggling" about important issues, instead, will state his position and keep to it.
4. Values accuracy in direct language and will speak very frankly.	4. Prefers indirect communication, avoids offending others or causing embarrassment.
5. Does not feel tied to conventions; is receptive to innovation.	5. Prefers to follow previously established conventions.

You can see from these descriptions of high context and low context negotiating styles that the way a person prefers to negotiate is a complex combination of attributes. This style will influence the way you and your client view each of the issues that affect your participation in a project. It will also affect how you and your client interact to come to an agreement. Most people will have a negotiating style that is somewhere in between the two extremes of high context and low context, and may include elements of both. As an adaptable consultant, how should you prepare when you are meeting a client for the first time? A good strategy is to be prepared for the rapport-building activities preferred by a client with a high context style. This is what Dr. Derr did in the positive version of her first meeting. If your client prefers to prolong the preliminary small talk, you can take this as a clue that he may prefer a high context style of negotiation. If he prefers to start discussing the project immediately, this might reflect a lower-context style of negotiation. You can then remain attuned to other clues about the way your client prefers to negotiate.

6.5 Identify the Issues

In Chapter 2, you learned about ten key issues in statistical consulting. They were introduced as key points of vulnerability in the consultant-client interaction. For this reason they now become the key points of negotiation in this chapter. As a reminder, the issues are repeated in Exhibit 6.2.

Exhibit 6.2 ***Issues of Negotiation in Statistical Consulting***

Issue # 1: What is your role?

Issue # 2: What are the roles of others on the project?

Issue # 3: How will communications be maintained?

Issue # 4: What are the deliverables?

Issue # 5: What are the deadlines?

Issue # 6: How will you be compensated for your participation?

Issue # 7: What are acceptable statistical practices?

Issue # 8: What are the ownership rights?

Issue # 9: What stipulations are there for security and confidentiality?

Issue # 10: When is your participation finished?

It is important to be able to identify which issues will affect your involvement in a specific project. Then you can focus your attention on coming to an agreement about them. As you learned in Chapter 2, dissatisfaction can often be traced back to unmet expectations that you and your client have concerning these key issues. You can help prevent dissatisfaction by developing clear agreements with you client. For example, consider the issues that Dr. Derr and Mr. Johnson covered in segment 5 of the video: They talked about options for the statistician's role (Issue #1) and the role of the client and his advisory committee (Issue #2). They agreed to send information to each other by email (Issue #3) about cost estimates, study design, and population size (Issue #4). Dr. Derr presented some options for cost (Issue #6), one of which involved a student intern. This intern would need to make use of the data (Issue #8). Mr. Johnson then expressed concern about confidentiality of the respondent's information (Issue #9). As this segment demonstrates, it may only take a few minutes to identify the key issues that will shape a negotiation.

For another brief description of a consulting experience that involves a number of issues, see Example 6.1:

Example 6.1

Nathan Thomas† works as a statistician in a medical center. He met with Josef Martin, the "go-between" for a physician. The physician, Dr. Frances Picardo, needed some analysis done for a research proposal. Josef was not entirely clear on what Nathan was supposed to do. He had to guess about some of the analyses. The result of having this go-between was that Nathan did a lot of analyses that were probably not necessary, and he was left wondering whether he answered the questions that Dr. Picardo really had. In the end, Nathan only billed for about half the time that he had worked on the project.

Example 6.1 illustrates the vulnerabilities inherent in statistical consulting. There were a number of issues that appear to have caused a problem: Nathan was not sure what he was supposed to do (Issue #4). Josef the "go-between", who acted as the representative of Dr. Picardo, appeared to have scrambled the communication (Issue #3). Nathan appeared to be unsure of the authority and responsibilities of the go-between in relation to Dr. Picardo (Issue #2). He was also unsure about how to charge for the project (Issue #6). Dr. Picardo may also have been dissatisfied with the outcome of the work. This dissatisfaction was likely to be over issues of timing (Issue #5), deliverables (Issue #4), and cost (Issue #6). This example gives the impression that the statistician, the physician and the go-between did not have a clear understanding of the circumstances governing their work together.

It is useful to look at examples of statistical consulting from a variety of fields that represent different issues, exchanges and outcomes. Several examples are shown below that will be used in subsequent exercises and illustrations in this chapter:

Example 6.2

Scott Lasser is a statistician at a biostatistical consulting unit. For three years, part of his salary support came from a research grant awarded to Dr. Satya Sethuramen, a scientist in the field of nutrition. She was preparing to conduct a

† The names and details of examples appearing in this chapter, unless otherwise indicated, are fictitious.

clinical trial of the effects of a specific vitamin on the health of nursing mothers and their infants. This study required that the study subjects be randomized into three groups that differed in the dosage of the vitamin to be given to the mother. The effect of the vitamin would be determined from periodic blood and milk samples from the mother, and blood samples from the infant, and other health measures. The study was expensive to conduct and the research budget was very limited. Based on some previous data in the literature, Scott estimated that Dr. Sethuramen would need about 30 mother-infant pairs in each of the three groups in order to detect a clinically important effect. Because of the limited budget and the cost of recruiting and enrolling mother-infant pairs, Dr. Sethuramen wanted Scott to analyze the data repeatedly as it accrued. She would then stop the study at the point at which statistical significance was obtained. Scott was very concerned about this request. The proposal to test data repeatedly while information was accruing violated his sense of what was good statistical practice. To back up his position, Scott did a literature review on the topic of interim analysis of data, and spoke with several experts. He was able to persuade Dr. Sethuramen that there were problems associated with this proposed strategy. They agreed instead that Scott would conduct a formal interim analysis at two intermediate stages of the trial.

Example 6.3

Michael Gregoriou works for the statistical support unit of a large engineering firm. One day he received a telephone call from Lawrence Nelson, an engineer who had what he believed to be a quick question about some data analysis on a statistical software package. Dr. Gregoriou was busy working on a project, but he provided a brief answer to the question. An hour later, Dr. Nelson called again with another question. Somewhat reluctantly, Michael provided another answer. A half-hour later, the engineer called again. This time Dr. Gregoriou answered very briefly with his annoyance clearly apparent in his voice. The engineer did not call again.

Example 6.4

Valerie Scranton is a statistician at an academic consulting unit. Dr. Ramon Guitterez, a professor at the university, asked her to help with a research study that had already been conducted. During the course of the experiment, a serious flaw had developed with the design, due to circumstances that were beyond anybody's control. It was far too expensive to repeat the experiment, which involved exposing animals to the zero gravity conditions of space flight. Dr. Scranton worked out a method that could be used to accommodate the flaw, and carried out the appropriate analyses. Dr. Guitterez developed a manuscript and asked Valerie to provide the language necessary to discuss the statistical methods and the limitations to inference. She wanted to be included as a co-author of this publication, but she provided the section on statistical methods without discussing this request with the scientist. She charged the unit's hourly rate for her time. The publication was accepted for publication, but Guitterez did not include Valerie as a co-author. She did not work with this professor again.

Example 6.5

In this example, we return to the story from Chapter 2, involving Neil Snowball, the statistician with his own consulting business. When we left the story in Chapter 2, Neil had been discussing a possible consulting contract with Jack Frost, a representative of the state's Internal Revenue System. In this episode, Neil has signed the contract to develop a statistical program to forecast certain tax revenues. In this contract, he has agreed to develop a forecasting model from data to be provided by the agency. He will then write a computer program to implement the model on the agency's computing system. The contract stipulates a deadline for the delivery and successful implementation of the program. Neil's payment will be a fixed amount of money to be paid over a period of three months. The contract also stipulates that he may publish the results of his statistical work only with permission of the state agency. Jack suggested informally that this permission would be forthcoming, as long as Neil masks the data appropriately.

We pick up the story again at the end of the project. The project took much more time than Neil expected because of the computer programming. He did not realize when he estimated his time that he would have to develop the forecasting program on the out of date programming language of the agency's computing system. The agency provided Neil with a "no-cost" extension of the deadline in order to complete the computer program. Neil found the process of building a forecasting model with the agency's data to be very interesting and rewarding, and he wrote up a manuscript for publication. The agency refused permission for him to submit this manuscript. For several years afterwards Neil consulted annually with this agency, helping them to update the program and interpret the forecasts. He charged the agency an hourly rate for this consulting. He also continued to press for permission to publish, but was not successful. Eventually Neil stopped consulting with the agency. He doubts that they can continue to implement the forecasting program without regular statistical support.

Example 6.6

Brenda Hollis is a statistician working in a government research unit. Brenda was asked to analyze some data from a research project that was a high priority for the agency. The agency scientist in charge of the project, Dr. Magda Landovska, was planning to present a paper on the results at a professional society meeting. Dr. Landovska was also going to be promoted soon to a position that had oversight over the statistics unit. Brenda's supervisor, Louis Chen, asked Brenda to set aside her other projects and make sure that the analysis was finished in time for the presentation. The data from the project was extensive and required a lot of manipulation prior to analysis.

Brenda met with Dr. Landovska and they identified the highest priority tasks that were needed for the presentation. They agreed to defer the remaining tasks to be done with lower priority after the presentation. Brenda enjoyed developing the statistical models for the analysis and felt comfortable with the judgment required to implement them. The rest of the work was fairly tedious, requiring considerable effort to produce a large number of graphs from which the Dr. Landovska would select the final graphs for the presentation. They agreed that after the presentation, Brenda would analyze the remainder of the data and then show Dr. Landovska how to develop the graphs from that point. Dr. Chen expressed his appreciation to Brenda for enabling Dr. Landovska to meet her deadline, and reassured Brenda that usually the agency scientists are expected to produce their own graphs. Dr. Landovska also expressed her appreciation and invited Dr. Chen to discuss the needs of the statistics unit with her.

Exercise 6.1

Review Examples 6.2 - 6.6. For each one, describe the key issues that shaped the interaction. Select from the ten key issues in statistical consulting listed in Exhibit 6.1. Add other issues if necessary.

6.6 Characterize the Positions Held by You and Your Client

Once you have identified the key issues in a negotiation, you can then characterize your and your client's positions on each issue. In segment 5 of the video, Dr. Derr informed Mr. Johnson that she could give some advice about the survey design at no charge. A special arrangement between their two units made this free advising possible. However, any further involvement of the statistics unit in the survey would have to be on a fee basis. Dr. Derr also stated a condition of further involvement: the statistics unit would need to participate in the development of the questionnaire. Mr. Johnson expressed interest in involving the statistics unit further, but he also expressed some conditions: The survey budget was limited, and he was concerned about the confidentiality of information from the survey.

Part of characterizing a position is finding out whether it might be modified by negotiation. It is doubtful that Mr. Johnson would agree to involving an intern in the survey unless he felt very certain that the privacy of individual respondents was secure. It is doubtful that Dr. Derr would agree to participate further in the survey unless she approved of the questionnaire. However, between the limits of what would be unacceptable to either of them there still appears to be room for negotiation.

Several factors determine the extent to which a position is modifiable, and therefore open to negotiation:

(1) ***The number of people and the amount of bureaucracy involved in modifying a position.*** For example, the hourly rates charged by Dr. Derr's unit were set through an annual administrative procedure. She might be able to get these rates changed for a specific project, but it would take a lot of effort to persuade all the people at different administrative levels to approve of this change. In contrast, Neil Snowball, the independent consultant in Example 6.7, was the only person involved in setting his rates. He could change them whenever he felt the circumstances required a change.

(2) ***The extent to which you or your client feels bound by convention to adopt a certain position.*** A client may tell you "This is just the way we do things here." A person with a high context style of negotiation may be most comfortable with agreements that follow the conventions of his discipline or of his work environment. For example, the physicians at the medical center in Example 6.1 have always regarded statisticians purely as "number crunchers" who do not need to be directly involved in the discussion of a project. Changing this perception may be a worthwhile effort, but it would probably require a long-term campaign to change the work culture. By contrast, Dr. Landovska in Example 6.6 asked about the needs of the statistics group and might be more receptive to innovations proposed by this group.

(3) The extent to which a position is defined by a deeply held belief or principle. You may be unwilling to do something that violates your sense of ethics or your beliefs about what constitutes good statistical practice. Scott Lasser, the statistician in Example 6.2, was very concerned about the multiple testing that Dr. Sethuramen wanted to do while the data was accruing. This violated his sense of what was a good statistical practice.

The most readily negotiated issues are those that you and your client can resolve by discussion between the two of you. Issues become more difficult to negotiate as they involve more people, more levels of organization, and run counter to more conventions and beliefs.

Exercise 6.2

Look through Examples 6.2 - 6.6 again. Find and describe the following: (1) A position that could be modified within the client-consultant relationship; (2) A position that would require effort outside the client-consultant relationship to modify; (3) A position that one of the parties held because of his or her beliefs or principles.

6.7 Achieve a "Win-Win" Outcome

A negotiation is successful when you and your client are aligned on the expectations for the key issues that affect the project. This is a "win-win" outcome in statistical consulting. The term "win-win" is often used in the field of negotiation to refer to the outcome in which all parties involved feel that their needs have been met (Risher and Ury, 1981). A "win-win" outcome in statistical consulting occurs when both you and your client understand and accept the basis for your work together, and each of you feels that the terms of the arrangement are fair and beneficial. The term "win-lose" is used to represent an outcome in which at least one of the parties feels defeated in the negotiation by the others. This type of "win-lose" outcome can erode the relationship between you and your client. In a "lose-lose" outcome, none of the parties has had their expectations met, and none are satisfied.

A story of a "win-win" negotiation is given in Example 6.7:

Example 6.7

Frank, the area manager of an industrial firm, was surrounded by his staff in an informal meeting. He explained the problem to Tom, the young statistician: "We called you in today to see if your so-called experimental design methods can help out with our yield problems. Anyway, Harry told us to try your stuff out. We don't have much of an expectation since our efforts have not been able to get the yield on the photoreceptors above 34%." Harry was another area manager in the materials section who had seen the power of statistical methods. Frank was a skeptic and was only humoring Harry.

"Let's look at the factors you have in mind that could influence yield" suggested Tom. "Well, we have these that we have looked at in the past, but I don't know if you can do anything with them any better than we have" replied Frank. "What exactly causes the defects that decrease the yield?" continued Tom. "Why does this matter to you? All we need to do is get the yield up to 50% - 60% to keep upper management off our backs. Well, it's surface defects. You know if there is a dimple on the surface, the paper will not make good contact and there will be a deletion. Can't have that you know" puffed Frank as he drew on his Lucky Strike.

Tom thought for a bit and summarized the new information. "So the overall goal is to keep upper management off your back by doubling the yield." "Right, if we can do that we'll all be heroes" echoed Gary, Frank's second in command. Tom continued, "Right now the objective is to find the relationship between the surface defects (dimples in particular) and these factors. How have you done your previous experiments?

Gary jumped in with enthusiasm. "We were very scientific, and systematic. We set up the evaporation chamber with all of the factors in a base condition and then made five modifications of each of the five factors moving each one all by itself so as not to get confusing results. I supervised all the trials myself and can testify that we only changed one thing at a time!" "What did you find?" asked Tom holding back his negative thoughts on the approach called 'one factor at a time' which he had learned was a very inefficient testing method. "Well, four of the factors had a bit of an effect, but we couldn't get anything above 34% when we used the best combination of them" confessed Gary. He went on, "Maybe you can work some magic with your statistical analysis on our data and get a better result. It's good data."

"I'm sure you did a good job of configuring the chamber, but I would like to suggest another set of trials that might do a better job. If you use the following set of trials (Tom began to set up a factorial design with four factors at two levels using the factors that Gary had indicated were influential in the one-factor-at-a-time tests) that are designed to look at not only the additive influences, but also the strange non-additive effects called interactions."

"We don't need to look at complex things like 'inneractions' or whatever you call them" brashed in Frank. "Let's hear him out, Frank, he may be on to something" cut in Gary who continued, intrigued by the runs Tom was putting down on the flip chart so everyone could see. "I see a few of the ones we already have run in Tom's list. We could just add a few more to complete the job." "I don't know if there are only a few more. Look at the number of runs he has outlined. He's got 16 runs. Gary you did it in only 6. See, I told Harry that statistics makes more work. This proves it," said Frank in a vindictive tone.

Tom replied playing his ace in the hole. "What if I told you we could look at all 5 factors with just the 16 runs I have outlined." Frank snapped back "But that 5th factor never did anything. But, I kind of like the idea of a freebee. Sure, I think it will make Harry happy. Let's compromise and include 'Tom's Factor' just for fun." The meeting ended. Tom built the ½ fraction, 16 run design and 'Tom's Factor' interacted with another factor. When the optimal conditions were tried, the yield went to 80%. Savings: $1 million per year. The compromise paid off. Frank was not completely converted to statistical experimental design, but he took full credit for the accomplishment.

Example 6.7 is a success story indeed! Tom managed to overcome Frank's skepticism, and the efficient design produced results that were used to specify optimal conditions. The revised process had a greatly improved yield. We can speculate that Frank had a "high-context" style of negotiation to the extent that he seemed to resist innovation and to prefer to do things the way they had always been done. Tom got Frank's attention by expressing the core problem from Frank's point of view. He may have worked with Gary and Harry on previous projects. They appreciated the value of statistical methods and in turn used their influence to persuade Frank. Tom also created good will by being diplomatic about their previous experiments. He held back his opinion about 'one factor at a time' designs, and instead praised the way they had configured the chamber. He also did not react to Frank's taunt that statisticians make more work. After the study was finished and successful, he graciously permitted Frank to take credit for the success to upper management. In doing so, he cultivated an important source of influence in his future work with this firm. In addition, he had an in-house example of the value of experimental design and the importance of interactions. He could then use this example to persuade other decision-makers in this firm!

In contrast to the success story in Example 6.7, "lose-lose" is the best description of the outcome of Example 6.1. Nathan, the statistician, appears to have done a lot of work, not really knowing if it was the work that he was supposed to do, and then he was not sure how he should charge for the work. He was not even sure that Dr. Picardo had her questions addressed. Not only was Nathan dissatisfied with the immediate interaction, but his actions may have perpetuated a pattern that will continue to be unsatisfactory. Why didn't he contact Dr. Picardo directly, find out what she wanted, and agree on appropriate charges? Perhaps some conventions at work supported what happened in this example: Physicians in this medical center may generally rely on assistants, or "go-betweens" to get much of their research done. However, Dr. Picardo may have been equally dissatisfied with the outcome: at the amount of time the work took, how much it cost, and the fact that some of her questions were still unanswered at the end. Both the statistician and the physician in Example 6.1 may have concluded this interaction with some unpleasant generalizations about each other's professions. Negotiating a win-win outcome in this environment will probably require changing some conventions to provide for more direct communication between the physician client and the statistician. This is an effort that goes beyond the immediate client-consultant relationship.

A "win-win" outcome promotes the development of a collaborative relationship between statistician and client. When all parties have their needs met and are satisfied with the basis of their working relationship, they are more likely to continue to work together. The other outcomes, "win-lose" and "lose-lose," lead to dissatisfaction, the withdrawal of communication, and the development of negative stereotypes and negative attributions by clients about consultants and by consultants about clients. For this reason, the focus in this chapter is on the cultivation of "win-win" outcomes in the negotiations between statisticians and clients.

Exercise 6.3

Refer again to Examples 6.2 - 6.6. For each example, do the following: (1) Decide whether the example represents a "win-win", "win-lose", "lose-lose" or some other outcome. Explain your reasoning. For the "lose" outcomes, do the following: (2) Describe what the statistician and client could have done to improve the outcome. (3) Speculate about improvements that will require work outside the immediate client-consultant relationship. Feel free to embellish the examples if that will help illustrate your point of view.

Exercise 6.4

Below is a description of a program that provides "free" statistical consulting. After reading the description, do the following: (1) Decide whether the example represents a "win-win", "win-lose", "lose-lose" or some other outcome. Explain your reasoning. (2) Identify all of the exchanges that take place among all of the parties involved. Do these exchanges seem fair to you? Discuss your viewpoints.

Art Lowe is a graduate student in a statistics program. He is taking a course in statistical consulting. As part of this course Art meets with graduate students from other departments, and provide advice about the design and analysis of their research studies. This "short-term" consulting is provided free of charge. However, in order to obtain this advice, student clients must fill out a form and schedule an appointment. The clients are also informed in advance about what to expect: the consultant will not provide answers in the first meeting, but will deliver a recommendation in the second meeting. The short term consulting is limited to these two meetings. Each time Art meets with a new client, he explains that he will typically spend about six hours developing a recommendation. He asks the client to focus on her highest priority questions. At the end of the first meeting, Art writes down the agreements that have evolved from the meeting about what he plans to do and what he would like the client to do. At Art's discretion he can schedule a third meeting with the client to follow up on his recommendations. If he wants to work further on a project, he needs to check with his instructor first. He usually gets permission, if it's a topic that he would like to learn more about.

Exercise 6.5

Find an experienced statistician who is willing to relate two consulting experiences to you: one example which he felt was a "win-win" experience and one example which was either a "lose-lose" or a "win-lose" experience. Ask him for his assessment of these experiences, and what he might have done to improve the "lose" outcomes. Then do the following: (1) Describe the experiences and the statistician's assessment. (2) Identify the key statistical consulting issues that shaped the negotiation. (3) Identify the position(s) held by the client and the consultant on the key issues. Characterize the extent to which each position was modifiable by negotiation. If a position was not readily modifiable, discuss whether this was due to bureaucracy, convention, or principles (or some combination of the three). (4) Speculate about what you might have done to improve the "lose" outcomes. Consider both improvements within the immediate client-consultant relationship and longer-term campaigns for improvement within the work environment.

6.8 Identify Fair Exchanges

In the examples and exercises in this chapter, you have read about many different types of exchanges. Some were among tangible items, such as money, salary support, reports and publications. Some items were intangible, such as good will, opportunities to learn, and the promise of future benefit. A "win-win" outcome in negotiation requires that all parties feel that the sum of the exchanges between them is fair. This can be a complex issue in statistical consulting. We must consider not only the tangible exchanges but also the intangible ones. Clients and consultants derive different benefits from their involvement in a project. Each person places a different value on these benefits, and part of searching for a "win-win" outcome involves making sure that the client's and consultant's values are compatible. Another complication comes from the fact that tangible and intangible items may sometimes be substituted for each other. Individual consultants and clients often feel very differently about the extent to which an intangible item, such as authorship, may substitute for a tangible one, such as money. This section will help you identify combinations of tangible and intangible benefits can lead to a fair exchange.

When you and your client are negotiating an exchange between tangible items, such as money paid for the analysis of data and the production of a report, the establishment of what is fair seems relatively straightforward. You can estimate the amount of time you will spend on a project. In order to establish a reasonable and competitive rate for your time, you can find out what other consultants are charging and make use of guidelines that others in similar situations have developed. Chapter 5 "Business Aspects of Consulting" in Boen and Zahn's book (1982) provides a good introduction to this topic. However, even a relatively direct exchange of tangible items requires careful consideration. Estimating the time that a project will take is not easy. It is far more typical to underestimate the amount of time it will take to complete a task than to overestimate it. For example, if you are working with data generated by a client, it may arrive in a format that takes far more time to work with than you had anticipated. Or, you may discover that the data does not fit the assumptions of the statistical model you had planned to use. In Example 6.6, Neil Snowball discovered that the programming task was going to take much more time than he had estimated. He had not asked about the programming language at the negotiation stage of the contract. In this example he did the additional programming work at no cost to the state agency, having obtained a no-cost extension to the deadline of the contract. Why did he not ask for additional funds? There are many possibilities: Perhaps he felt it was his error in not asking about the programming language in the first place. Perhaps the agency had made it clear that there was no more money available for the contract. Perhaps he wanted to build some good will with the agency so that they would approve the publication of his manuscript and consider him for future contracts. Even in an exchange of tangible goods and currency, there are usually additional exchanges among intangible items that influence the outcome.

Intangible benefits play an important part in statistical consulting. In Example 6.6, most of the exchanges were of intangible items. Brenda Hollis, the statistician, earned the appreciation from her supervisor for stepping in and doing the tedious work required to support a project that had a high profile within the agency. All parties involved viewed the exchanges as fair: Brenda produced the graphs for the presentation in exchange for appreciation from her supervisor and the promise of future benefit to the statistics unit as a whole. Dr. Landovska, the agency scientist, obtained the statistical support she needed to deliver an important presentation in exchange for an obligation towards the statistics unit. Louis Chen, the statistics supervisor, set aside the other priorities of the statistics unit in exchange for the prospect of future benefit from the client who would soon be promoted to a position of influence over the unit. Although financial support for Brenda's time on the project may have also been involved, the key consulting issues are the timing and the priority of the activities that were negotiated through these intangible exchanges.

What intangible benefits does a statistician obtain from working with a client? It can be very rewarding to know that you have accomplished a challenging task. Most statisticians like to solve problems. Knowing that you have developed an efficient sampling plan, or a study design that addresses all of the investigator's goals, or an analytical model that accommodates all of the quirks of the acquired data, can be rewarding in itself. Your satisfied client can lead you to other prospective clients. This will enhance your reputation among clients and increase the number and types of projects from which you can select. Your work can also lead to increased recognition from your own peers in statistics. Most statisticians like to continue learning and mastering new skills, and so the opportunity to learn through the challenges of a client's project is another benefit. You may enjoy the feeling of belonging to a project team and being a part of an effort that is larger than yourself. You may also feel rewarded by working on projects that you feel are important.

Intangible items greatly influence the outcome of a negotiation between statisticians and clients. However, their value is difficult to quantify. You can often get a feel for how much value a person places on an intangible item by looking at the decisions he made concerning that item. For example, Scott Lasser, the statistician in Example 6.2, was very concerned about Dr. Sethuramen's plan to analyze clinical data repeatedly as it accrued, presumably with no adjustment to the error rate for these repeated tests. Scott took the time to review literature on the interim analysis of data. He also worked to persuade Dr. Sethuramen to change her protocol, despite her concern about cost. He clearly placed a high value on what he believed to be good statistical practices. A more stringent test of this value would have occurred if Dr. Sethuramen had refused to follow his recommendation. Would Scott have continued to work on the project? What would you have done? In Exercise 6.6, you will have the opportunity to examine the relative value you place on different intangible benefits.

What intangible benefits do clients gain from their work with statisticians? We would like to think there are many such benefits. A statistician's involvement in a project should give a client confidence that the goals of her project will be accomplished. A statistician should increase the value and improve the quality of the project. This should lead to increased recognition by the client's peers. Many clients also feel rewarded when they learn more about statistics from you. They may prefer to understand the rationale behind important decisions that affect their projects: the choice of sample size, study design, analysis strategy, or the interpretation of results. By including you on her project team, a client should also benefit from your perspectives and problem-solving skills. This should give the project team a broader scope and more inspiration for future projects. Of course, we should really ask clients directly about such an important topic. In Exercise 6.7 you will have the opportunity to interview a client about the intangible benefits she obtains from working with a statistician.

When you and your client first meet, it is helpful to remember that each of you derives different intangible benefits from the project. The relative value that each of you places on these benefits will influence the decisions that you make. Part of working towards a "win-win" outcome will be to make sure that your and your client's values are compatible. We can speculate that Tom, the statistician in Example 6.7, valued his continuing relationship with this firm. Frank, the skeptical manager, wanted senior management to see him as an effective manager who could solve problems and save the company money. Tom was able to make sure that Frank's needs were met in a way that also improved his position with the firm. In contrast, the values of the Neil Snowball and the state agency in Example 6.5 appeared to be in conflict with each other. Neil appeared to place a high value on publishing his work on the forecasting model. We can speculate that agency personnel may have perceived no benefit to the publication that would offset the potential risks, no matter how small, in disclosing information about their forecasts. In this situation the agency had the final authority and Jack was not able to publish the manuscript. This was a "win-lose" outcome, or, if you consider the fact that after a period of time Jack stopped supporting the forecasting model, it was a "lose-lose" outcome. We can speculate that Jack stopped working on this project when he found a more rewarding opportunity. In Exercise 6.8, you and the client you are interviewing will have the opportunity to examine the way you each value different intangible benefits and decide whether they are compatible. This should give you a model for how to discuss and evaluate these intangible benefits with your actual clients.

Intangible items may also serve as substitutes for tangible items. For example, if a client does not have much money to offer, he may offer some intangible benefit instead, such as the reward of working on an important project. Many statisticians do volunteer their time for projects that support their values and beliefs. Within the professions of law and medicine, a certain amount of *pro bono* work is expected as part of a practitioner's obligation to society. You may work for no tangible reward if you place a high value on the intangible benefits that derive from it. The extent to which you feel that an intangible benefit can substitute for a tangible item will be reflected in the decisions you make. In Exercise 6.6 you will have the opportunity to examine your own views by considering hypothetical projects that involve this type of substitution.

Statisticians and clients are likely to view the substitution of tangible and intangible items differently. A common point of disagreement concerns whether or not authorship can be substituted for payment. In many statistical careers, evidence of scholarly activity is an important part of evaluation and advancement. Many statisticians feel that authorship should be given to everyone who makes a substantial intellectual contribution to a project, and that this benefit is entirely separate from financial compensation. In contrast, many clients will offer a statistician co-authorship on a paper as a substitute for money. They may do this to stretch an inadequate project budget or to follow already established conventions in their field. They may provide funding in lieu of authorship because they are not aware of or don't recognize the intellectual effort involved in the statistical contribution. In Example 6.4, Dr. Scranton, the statistician, probably felt that she should have been included as a co-author because of her important contribution to the research effort. Without her help, Dr. Guitterez would not have known how to accommodate the unanticipated flaw in the study design. Valerie felt that her statistical problem-solving ability enabled the data analysis to move forward and was an additional, intangible contribution that was not covered in the hourly charge for her time. She felt that being a co-author on the manuscript would be a more appropriate exchange for her intellectual contribution to the project. In contrast, Dr. Guitterez must have felt that the hourly charge for the statistician's time was sufficient compensation. Because they did not work together on future projects, we can guess that one or both of them were dissatisfied with the exchanges. In Exercise 6.7 you will have the opportunity to find out the extent to which the client you are interviewing would allocate funds from her project's total budget in order to obtain different intangible benefits from the statistician. This is a way of finding out the client's perspective on substituting tangible items for intangible ones.

Intangible items have a great influence on the outcome of a consulting project. If you are new to a job, or you have not worked with a certain client before, it will probably take you some time before you understand the typical exchanges and their relative value in the workplace. The less you know about a client and your work environment, the more risky it is to make assumptions about exchanges. Your client may have a very different view about the relative value of the items that you and he may be exchanging. He also may feel differently from you about the extent to which one item may be substituted for another. If you would like to learn more about exchanges of intangible items in the workplace, an excellent resource is the book *Influence Without Authority* by Cohen and Bradford (1991). In this book, Cohen and Bradford explain the theory of reciprocal exchange and show how to develop win-win outcomes by developing strategies for the exchange of intangible items.

Exercise 6.6

Shown on the chart below are some intangible benefits that a statistician can obtain through working with a client.

Intangible benefits to statisticians obtained from working with clients. (Feel free to add other intangible benefits to this list.)	***A. Rank the importance of each benefit to you with 1 = the most important benefit.***	***B. What percentage of your financial compensation on a project would you give up in return for obtaining this benefit?***
1. The reward of accomplishing a challenging task.		
2. Increased recognition from peers in statistics.		
3. Increased number of contacts with other clients.		
4. The feeling of being included on a team.		
5. The feeling of being involved in something important.		
6. The reward of learning something new.		
7.		
8.		
9.		
10.		

Feel free to add other intangible benefits to this list before going on to the following activities:

A. For each of the benefits on your list, think about what their relative value is to you at this point in your career. Produce a ranking of these benefits with 1 = the most important benefit. Ties are okay, and "Not applicable" is also an acceptable response.

B. For each of the items on your list, consider the extent to which you would be willing to substitute it for a tangible benefit such as financial compensation. Indicate the percentage of the total financial compensation of a project that you would be willing to give up in order to obtain each benefit.

Write a brief report that summarizes the results of your comparisons.

For the next two exercises, find a client who will be willing to discuss some of the intangible aspects of negotiation with you.

Exercise 6.7

Shown on the chart below are some intangible benefits that a client can obtain through working with a statistician.

Intangible benefits to clients obtained from working with statisticians. (Feel free to add other intangible benefits to this list.)	*A. Rank the importance of each benefit to you with 1 = the most important benefit.*	*B. What percentage of a typical project's budget would you allocate to statistics in order to obtain this benefit?*
1. Increased confidence that a project's goals will be accomplished.		
2. Increased recognition from the client's peers.		
3. Increased range of capabilities within the project team.		
4. The stimulation of new ideas for future projects.		
5. The reward of learning something new about statistics.		
6.		
7.		
8.		
Ties are OK. Some benefits may be "Not Applicable" to your situation.		

Show this list to the client you are working with on this exercise. Expand it to include other intangible benefits that she values in her work with statisticians.

Then ask her to consider the following questions from the perspective of a typical project in her field:

A. Ask her to rank each of the intangible benefits that statisticians can provide. Ties are okay, and "Not applicable" is also an acceptable response. This should give you an idea of the relative value of each of these intangible benefits to this client.

B. For each of the intangible benefits on this list, find out the extent to which the client would be willing to allocate the financial resources of a typical project to statistics in order to obtain this benefit.

Write a brief report that summarizes the results of your discussion.

Exercise 6.8

Consider a typical project in the field of the client you are interviewing. Decide whether the two of you could negotiate a fair exchange among intangible benefits, including the possible substitution of intangible benefits for financial compensation. In order to make this decision, do the following:

A. Show the client the ranked list of intangible benefits that you developed in Exercise 6.6. Compare the 2-3 items that you value the most with the 2-3 intangible items that the client values the most (from Exercise 6.7). Discuss the relative compatibility of your lists.

B. Show the client the substitutions you would be willing to make between money and intangible benefits (from Exercise 6.6). Compare this with the resource allocation that the client would be willing to make to obtain intangible benefits from the statistician from Exercise 6.7. Discuss the relative compatibility of your substitutions.

Write a brief report that summarizes the results of your evaluation.

6.9 Conduct the Negotiations

In the previous sections of this chapter, you have gained an understanding of the issues that are important in statistical consulting. You have learned how to characterize the extent to which issues are negotiable. You have identified the exchanges between consultant and client in a variety of examples. You have seen how a fair exchange is forged from tangible and intangible items based on your and your client's perceptions of the value and substitutability of these items. Armed with all of this knowledge, how do you go about actually conducting a negotiation with your client? In Exercise 6.8, you and the client you interviewed had the opportunity to discuss tangible and intangible benefits, their relative value and the extent to which they may be substituted for each other. From this discussion you decided whether the two of you could negotiate a "win-win" outcome to your participation in a hypothetical project. This exercise gave you some practice in negotiation. As with most of the communication skills in this book, the practice that comes from experience will be your best teacher. In this section, you will find some guidelines about the influence of communication style and negotiation style on the way you conduct a negotiation.

In Chapter 4, you read about four dimensions of communication style that were especially important in statistical consulting. You covered phasing and sequencing in Chapter 4, and sequencing and specificity in Chapter 5. The remaining dimension, objectivity, affects the way you and your client prefer to conduct a negotiation. Objectivity refers to the way that people use language to express their ideas. As with the other dimensions, we can think of objectivity as a continuum of preferences between two opposing styles. On the one extreme is someone who prefers to use precise spoken and written language to convey his meaning directly. An example of this type of direct communication is in segment 5 of the video. In this segment, Dr. Derr brought up the issues of compensation, the use of information, and other stipulations about her role directly. On the other extreme of objectivity is someone who prefers to be indirect, relying on context and inference to get his ideas across. For example, Valerie Scranton, the statistician in Example 6.4, appeared to prefer this more indirect style. She did not approach Dr. Guitterez directly about being included as a co-author on the publication. Instead, she appears to have relied on indirect inferences and other conventions about the circumstances under which a statistician is invited to be a co-author on a paper. Unfortunately for Dr. Scranton, Dr. Guitterez did not appear to follow the same conventions or to pick up on the indirect communication about this wish.

There is no right or wrong level of objectivity in statistical consulting. You can achieve the goal of clear communication regardless of the level of directness with which you prefer to negotiate. However, the outcome of Example 6.4 provides a note of caution about indirect communication: Since statistical consulting is interdisciplinary, it is very likely that you and your client will not interpret indirect communication in exactly the same way. Misunderstandings arising from unclear communication can then evolve into unsatisfactory outcomes.

As with the other dimensions of communication style, problems can also arise when you and your client have different preferences for objectivity. The person who discusses negotiation issues directly can seem aggressive and demanding to someone who prefers to use more indirect language. Follow-up written summaries can also seem intrusive and controlling. The person who prefers indirect negotiation can appear indecisive and even duplicitous to someone who prefers to have a direct discussion about these issues. The fact that this person may not produce a written summary can further reinforce this negative impression. These negative attributions, caused by differences in style, can interfere with the process of achieving a clear understanding between the client and the consultant. Without this clarity, it is less likely that the negotiations will lead to a satisfactory, or "win-win" outcome.

You can recognize a difference in preference for objectivity between you and your client in the same way that you have learned to identify differences in other dimensions of communication style. By monitoring your client's responses and your own reactions, you can form an impression about whether you and your client are similar or different about how directly you prefer to discuss issues. In segment 5 of the video, Mr. Johnson appeared to respond positively to Dr. Derr's direct discussion. If he had preferred a more indirect route to the negotiations, he might have done something to change the course of the conversation.

If you prefer to discuss issues directly and you encounter some resistance, or if you begin to form a negative impression that the client is not being open with you, stop and consider that the client may simply prefer a more indirect style. Likewise, if you begin to form a negative impression that the client is too aggressive and demanding, stop and consider that his preferred style may be more direct than yours. By paying attention to the cues of the conversation, you can overcome your negative impressions and reactions that can cause the negotiations to go badly. You may be able to respond to these conversational cues by modifying your own level of directness so that it is a closer match to the client's. By developing a variety of strategies you can obtain clarity in your negotiations with your clients. Below are some suggested strategies for statisticians and clients with different combinations of preference for objectivity:

1. **Both statistician and client prefer a direct communication style.** This can be a compatible combination. Both of you are likely to seek clarity through direct discussion. It should not be difficult for you to come to agreements and then exchange written summaries of these agreements.

2. **Both statistician and client prefer an indirect communication style.** This can also be a compatible combination. However, you must keep in mind the potential for differences in the way you interpret each other's inferences. It may help you to find out more about the client's background in order to understand what her priorities and expectations are. If you prefer an indirect style yourself, you are probably adept at the skills of intuition that will help you to interpret your client's indirect communication. Once you learn more about the client's way of expressing herself, you are likely to be able to establish clear agreements in the indirect style that is comfortable for both of you. Neither of you may be inclined to summarize your agreements in writing. However, you are each likely to adhere to other conventions that provide the necessary follow-through without a written agreement.

3. **The statistician prefers a more direct communication style than does the client.** It can be challenging to achieve clarity with this combination. Because you expect all of the issues to be discussed openly, you may miss some of the more indirect messages that the client is sending to you. You may unintentionally offend the client by your own style that the client can interpret as overly argumentative or intimidating. Consider explaining your style to your client. Since you prefer direct communication, this should not be too difficult. You can let the client know that you have a hard time understanding something unless it is put directly into words. Be sure to paraphrase the client's comments to make sure that you are interpreting them correctly. Explain that it helps you understand the agreements that you and the client have made if you summarize them in writing. At the same time, try to become more attuned to the client's indirect communication style. There may be as much meaning in what she does not say as in what she does. Be careful with your own words and body language, as the client may overinterpret them.

Learn as much as you can about the client's work environment and what conventions and priorities she may have that influence her negotiations.

4. **The statistician prefers a less direct communication style than does the client.** This is probably the most challenging combination of styles with which to achieve clarity in negotiations. Just by recognizing the difference between your two styles, you will have made progress in warding off the negative attributions that can derail your communication. Keep in mind that your client may not pick up on the indirect signals that you send to him. Try not to be offended by his more direct style. It may not be easy for you to discuss these differences in communication style. However, do attempt to find ways to be more direct with this client while still maintaining your own comfort level.

The dimension of objectivity in communication can provide you with an insight into the more complex set of attributes that characterize a person's preferred negotiating style. The client's level of directness in his immediate conversation with you can suggest what his preferred negotiation style is likely to be. If someone has an indirect communication style you can guess that he may have other attributes of a high context style of negotiation. You can then stay attuned to other clues about these attributes: positions he may hold because of conventions in his discipline or workplace, unwillingness to haggle about positions he holds out of belief or principle, a reluctance to argue, a preference for developing a personal relationship before negotiating. In order to promote a "win-win" outcome with this client it will help to address this high context style. For example, you can increase your acquaintance with this client outside the context of the project. Find a way to serve on a committee or task force with him. Or, create a social or business-social occasion that includes this client. You can also provide him with examples of your previous work that illustrates the conventions and principles that you follow. Because he respects conventions and principles, he may respond positively to this way of expressing your preferences.

If your immediate conversation with a client suggests that she may prefer a low-context style of negotiation, then you can look for clues about other attributes associated with this style: willingness to debate about most issues, tendency to state a position on an issue that is more extreme than the one she actually feels, willingness to "haggle" and compromise on many issues, willingness to consider ideas that are not part of the conventions of her workplace or discipline, and less interest in developing a personal acquaintance with you as a prelude to negotiations. To address this low context style you can focus your energies on the immediate discussion and regard it primarily as a problem-solving exercise. Even though the client may sound very definite about a position, if you challenge it you may discover that she is willing to modify it towards a more central position.

Negotiating a win-win outcome is a complex and interesting part of statistical consulting. Mastery in this area will require practice. Below is a summary of the recommendations in this chapter for improving the communication skills used in negotiations:

- Be sure to put negotiation on the agenda.

- Prepare for a high context style of negotiation. Then seek clues from the conversation that identify the client's preferred style of negotiation. Adapt the discussion accordingly.

- Identify the key issues that will affect your participation in the client's project.

- Identify your and your client's positions on these issues. Characterize them by the extent to which these positions can be modified. Find out why a position is not likely to be modified.

- Characterize the relative value of intangible benefits of the project to you and to the client. Find out the extent to which they may be substituted for tangible benefits. Evaluate whether the exchanges involving intangible benefits are compatible.

- Identify and adapt to your client's preferred level of objectivity in discussing issues.

- Identify and adapt to your client's preferred negotiating style.

- Work towards a win-win outcome!

You will know that your skills in negotiation are improving when you are able to negotiate "win-win" outcomes with more complex sets of exchanges and with clients who represent a broader range of styles. At the same time, you can learn to recognize situations that are very unlikely to lead to a "win-win" outcome. For example, if you and your client hold positions that are not compatible and not readily modified, then it is unlikely that the two of you can work together and be satisfied with the outcome. Similarly, if you and your client have incompatible views about the relative value of different intangible benefits and the extent to which they may be substituted for tangible benefits, you are unlikely to find a satisfactory exchange. This insight will help you to focus your efforts where they are most likely to lead to rewarding outcomes.

Exercise 6.9

Refer to Example 6.2. Put yourself in the place of the statistician. Describe how you would you go about persuading this client to adopt your recommendations if: (a) the client prefers direct communication; (b) the client prefers indirect communication. Be sure to embellish the example as needed to illustrate your ideas.

Exercise 6.10

Refer to Example 6.6. Put yourself in the place of the statistics supervisor. How would you go about negotiating the agreements with this client if: (a) the client has a high context style of negotiation; (b) the client has a low context style of negotiation. Be sure to embellish the example as needed to illustrate your ideas.

6.10 Suggestions for Group Discussion

1. Students are instructed to watch segment 5 of the video before reading through the rest of the material in this chapter. You can introduce this chapter by showing this segment in class.

2. If you are making use of other recorded statistical consultations in your course, you may be able to identify negotiation activities in portions of these tapes. You can have students study these negotiation activities, identify issues and characterize positions and exchanges in the same way as was done with segment 5 and with the examples in this chapter. You can also ask students to speculate about the preferred communication style and negotiation style used by each party on these supplemental tapes.

3. You may be fortunate to be able to work with a group whose members represent a range across the dimension of objectivity in communication style and a mix of the attributes found in high and low context styles of negotiation. Take advantage of this diversity in your discussions about strategies in negotiation.

4. Exercise 6.5 asks students to obtain consulting examples (one with a "win-win" outcome and one with a "lose-lose" or "win-lose" outcome) from an experienced statistician. If you are working with a group of experienced statisticians, they can draw these examples from their own experiences. If you have access to a list of participants in advance of the course, you can contact them and ask them to bring these examples in with them. If you are working with inexperienced students, you will need to arrange for them to interview more experienced statisticians. This could be done in person or electronically.

5. Exercises 6.6 - 6.8 are complex and are designed to provide a forum for discussing the intangible benefits of the client-consultant relationship. There are no right or wrong answers. Students may have mixed success in achieving the endpoints requested by each exercise. Make sure they understand that it may not always be possible to achieve these endpoints. Encourage them to report on the process of working through each exercise and to describe any difficulties that they encountered along the way. These exercises should provide an opportunity for you to expose the class to the variety of viewpoints that exist among consultants and clients. Exercises 6.7 and 6.8 require work with a client. Although it would be helpful for students to speak with clients in person, you could probably also arrange these discussions to take place electronically. You could also make use of this set of exercises to organize a panel discussion involving experienced consultants and clients.

6. You may also want to organize a discussion of the business aspects of consulting. If your class includes experienced consultants, you can do this within the class. If none are experienced, invite an independent consultant to speak to the class. You can include a discussion of the different formal written agreements that involve statisticians, as described in the next teaching tip.

7. As a supplementary exercise you can ask different members of the class to bring in examples of written agreements that address some of the key issues in statistical consulting. Examples of these written agreements could include formal contracts, descriptions of policy, memoranda, letters or meeting notes. Across the class as a whole, attempt to obtain agreements from a variety of work environments. Each person who contributes an example can explain what the agreement stipulates about different key issues in statistical consulting. If you are working with an experienced group of statisticians and can contact them prior to the short course or workshop, have them bring an example from their workplace with them. If you are working with inexperienced students, have them contact an experienced statistician to obtain an example. Another way to organize this is to have an experienced statistician talk to the class and bring in several examples. This could be part of the session on business aspects of consulting. You can also collect some examples to round out the discussion.

6.11 Resources

Resources cited in this chapter:

Boen, J.R. and Zahn, D.A. (1982), *The Human Side of Statistical Consulting,* Belmont, CA: Lifetime Learning Publications.

Cohen, A.R. and Bradford, D.L. (1991), *Influence Without Authority*, NY: John Wiley and Sons.

Cohen, R. (1991), *Negotiating Across Cultures,* Washington DC: United States Institute of Peace Press.

Fisher, R. and Ury, W. (1981), *Getting to Yes,* Boston, MA: Houghton-Mifflin.

Other resources on negotiation:

Albrecht, K. and Albrecht, S. (1993), *Added Value Negotiating*, Homewood, IL: Business One Irwin.

Cohen, H. (1980), *You Can Negotiate Anything,* NY: Bantam Books.

Gray, B. (1989), *Collaborating: Finding Common Ground for Multiparty Problems,* San Francisco: Jossey-Bass Publishers.

Karras, C. (1974), *Give and Take,* NY: Crowell.

Lewicki, R.J. and. Litterer, J.A. (1985), *Negotiation*, Homewood, IL: Irwin.

7

COMMUNICATING ABOUT STATISTICS

7.1 Introduction

We finally come to the point in statistical consulting where you are ready to communicate some statistical information to your client. At this point, you have obtained the information you need, you have thought about it and worked on it. In terms of the model of statistical consulting we have been working with (Exhibit 7.1), you are translating your statistical solution to the client so that he may apply it to his project.

Exhibit 7.1 ***The Process of Statistical Consulting***

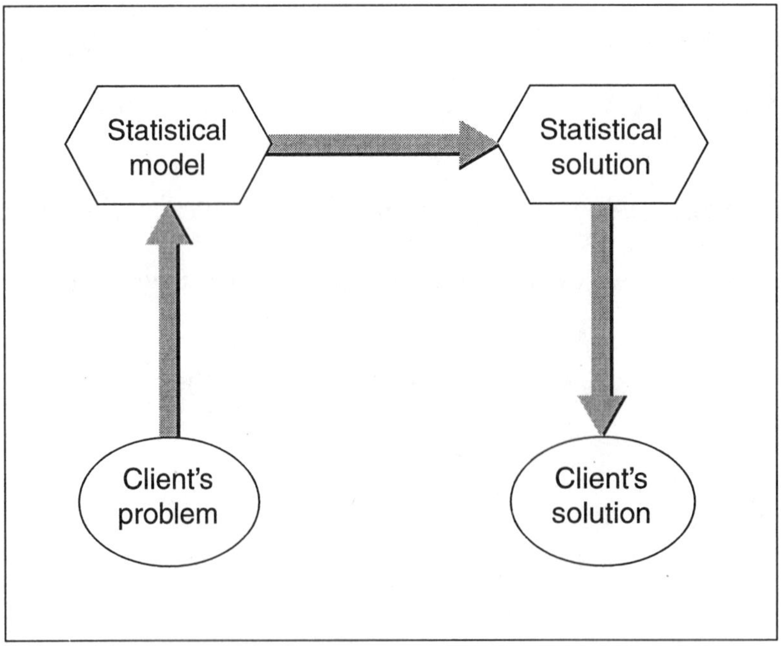

The way you communicate statistical information to your client takes many forms: A conversation, a presentation, graphs and tables, email, a technical report are just some of the possibilities. And, as statistical software becomes more and more user friendly, you may be spending more time showing clients how to do their own statistics. Margaret Nemeth (1998)

discussed the changes that user friendly statistical software has caused to the way she communicates about statistics:

> At Monsanto, researchers have access to JMP®. With this easy availability of software, researchers are doing more and more of their own statistical analyses. Thus, we must change how we interact with our clients. In the past, I designed experiments for clients, analyzed the data, discussed the results with the client, and wrote a report. Now I spend a portion of my time talking to teams about "statistical thinking" and teaching them how to use statistical software appropriately. When a team indicates that its members will be using JMP®, I have them send me representative data, and my presentation is geared toward analyzing data of this type. We also discuss assumptions, hypotheses, estimation, and so forth, but everything is in terms of their data and very few symbols and very little statistical "jargon" is used. I do emphasize when they should contact me and the limitations on what they have learned. I find this interaction more fulfilling, and it certainly beats sitting at a computer cranking out output.[1]

You are communicating effectively when your client understands you and can apply the information to the goals of her study and the decisions she needs to make. This is challenging because statistics is a difficult area for many people. Although you probably chose the field of statistics because of your interest and aptitude, your client probably has a very different set of abilities and interests from yours. You and your client may also possess different learning styles that will affect how each of you organizes and processes information. It will be up to you to recognize and adapt to these differences so that you may communicate statistical information clearly and persuasively.

7.2 Learning Outcomes

- Describe how to recognize and adapt to your client's preferences in learning.
- Identify effective methods for communicating statistical information in conversation, presentation, and in writing.

---

[1] From "Discussion" by Margaret A. Nemeth, p. 206.

7.3 Identify Your Audience and Your Purpose

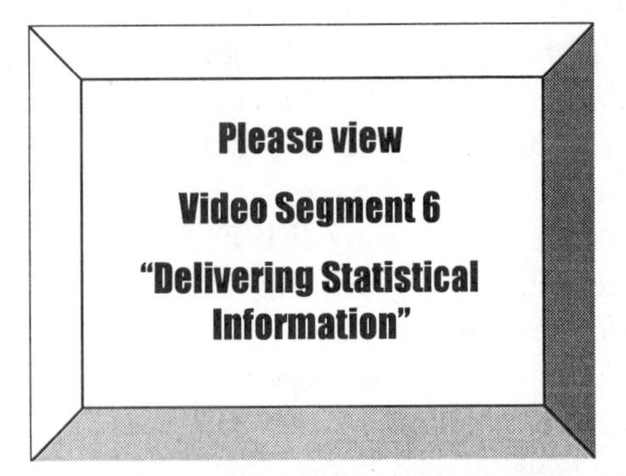

What happens when a statistician presents statistical information without considering her audience? If you can, take a look at segment 6 of the video now before reading any further in this chapter. In this segment Dr. Derr is presenting a sampling plan for the survey that Mr. Johnson would like to conduct. As you view the segment, focus on the way that Dr. Derr discusses her recommendations with Mr. Johnson. Do you think that Dr. Derr accomplished her purpose with this presentation? Well, that depends on what her purpose was. If she simply intended to present the information, she certainly did that. However, if she wanted to communicate the information to Mr. Johnson in such a way that he understood it and was able to make decisions on the basis of that information, then there is much more doubt about whether she accomplished this purpose.

As a consultant, you will often be communicating about statistics to a variety of audiences and in a variety of circumstances. To prepare for these situations, it is a good idea to consider who your audience is and what the purpose of your communication is. Here are some examples of typical situations in which a statistician needs to communicate statistical information:

Example 7.1

Jean Reina, the graduate student statistician in the WEST study, (see Example 5.4) wrote a technical report that summarized the statistical methods and results of the "virtual twins" assessment of the WEST program. The technical report had several purposes: (1) to summarize the statistical findings of the study; (2) to serve as a draft of the statistical portions of the manuscript that would be developed from the study; (3) to provide a permanent record of the details of the statistical methodology and the computer work. The body of the technical report was written for the director of the WEST study. In a separate technical appendix, Jean included details of the search of the data warehouse, the matching procedure, the database that she created and the statistical analysis. This appendix was addressed to a statistical reader.

Example 7.2

Russ Moodie, the head statistician for the FOOD study, (see Example 4.6) delivered a presentation to the steering committee. The purpose of the presentation was to provide members of the steering committee with information about the statistical power of the proposed multi-center study. This information would provide the statistical input into the decision that this group needed to make about the sample size for the study. Russ provided handouts of the transparencies that he used to illustrate his talk. These transparencies included power curves generated for the range of sample sizes under consideration.

Example 7.3

In the continuing video story of the UMS survey, Dr. Derr submitted a notebook to Mr. Johnson filled with tables that summarized the responses from the respondents of the survey to each question. The purposes of the notebook were: (1) to provide Mr. Johnson with some preliminary summaries of the survey, (2) to get his input about how to calculate the percentages for the executive summary and (3) to get his input about which information to include in the executive summary. The notebook was 100 pages long. The tables were produced after the data had been entered and cleaned. They were generated by a statistical software package that provided not only the summary statistics across all respondents for each question, but also cross-tabulated the responses to each question by the origin (domestic or international) of the respondent. As you will learn in segment 8 of the video, Mr. Johnson initially had a highly negative reaction to the notebook. However, once Mr. Johnson had provided his feedback, Dr. Derr developed a 10-page report. The purposes of the report were: (1) to provide information that the UMS committee could use to make decisions about possible improvements to their process, and (2) to provide a document that could be circulated to the committee's "stakeholders", that is, those with an interest in the issues of customer satisfaction with the medical services. Dr. Derr also prepared a separate technical report that included details of the sampling plan, the database, a summary of the response to the survey, and details about the decisions made during the process of the analysis.

Example 7.4

James High, the applied statistician at the telecommunications company, (see Example 4.4) set up a telephone conference call. Participating in the call were the members of the evaluation team, including statisticians, the survey research specialist, and representatives of the marketing group. The purpose of the conference call was to come to an agreement concerning the methodology to be used to evaluate the new product. Prior to the call, James had faxed all of the participants a table that summarized the advantages and disadvantages of each of the approaches that they were considering.

Example 7.5

Elaine Walker†, a statistician for the Food and Drug Administration, participated in the evaluation of a device used in veterinary practice. The lead investigator was Vincent Cantrell, an analytical chemist who was in charge of assessing the device. Elaine and Vincent worked together to design the study which was then implemented in Dr. Camtrell's laboratory. After the data had been collected and analyzed, Elaine wrote the statistical portions of the official report that summarized the evaluation of the measurement error of the device. Vincent wrote the remaining portions of the report and led the presentation to the review committee. At this meeting, Elaine presented the statistical portions of the study. The committee was made up of scientists, veterinarians and administrators from several divisions and was responsible for making a final decision about the device.

These examples show statisticians communicating in conversations, in more formal oral presentations, and in written material that ranged from highly technical statistical explications to very non-technical summaries and recommendations. The challenge in statistical consulting is to be able to communicate statistical information clearly enough in a variety of modalities so that the people you are working with understand it and are able to incorporate it wisely into their decision-making process.

7.4 Learn from Statistics Teachers

An effective statistical consultant can explain statistical concepts clearly. How can you improve the way you communicate statistical information to your clients? A good place to look for improved methods is the teaching profession. Although your role as a consultant is different from the role of a teacher, there are some useful similarities. For example, just as a classroom is made up of students with diverse backgrounds, your clients are also likely to vary in terms of their level of statistical knowledge, the extent to which they want to learn more about statistics, and the ways in which they learn best about statistics. As a consultant, you will probably have much less time with your client than a teacher does in a traditional statistics course. In fact, your role of consultant is probably more similar to that of an instructor of a short course or a workshop taught to non-statisticians. For this reason, it is important for you to choose methods of presenting statistical concepts that have a high impact when delivered to a non-statistical audience. The field of educational research can provide you with some ideas about how to improve the clarity of your statistical presentations.

† The names and details of examples appearing in this chapter, unless otherwise indicated, are fictitious.

An emerging theme in the field of educational research is the diversity of strategies people use to take in and organize information. Several theories of learning provide insight about how to teach statistics. For example, Kolb's theory of experiential learning has inspired teachers to include more active learning in statistics courses (Kolb, 1984). Students appear to learn statistics better when they are involved in projects and other activities. This is in comparison with "passive learning," or just sitting and listening to lectures. When students participate in problem-solving activities with real data in which they have a genuine interest, they appear to have better retention of what they have learned. They also appear better able to apply statistical thinking to new problems they encounter outside of the classroom (Cobb, 1991). This is good news for statistical consultants. Your client is involved in an actual problem in which he has a genuine interest. He should be in an excellent position to learn about statistics. All he needs is a good teacher!

In segment 6, Dr. Derr made use of passive learning techniques with Mr. Johnson. He sat in his chair while she delivered a lecture to him from across her desk. Perhaps you observed Mr. Johnson's body language and facial expressions during Dr. Derr's lecture. It is doubtful that he retained much from this lecture after he left the meeting or that he was able to apply it to his survey.

Another theory of learning that is relevant to statistical education is "whole-brain theory." Whole brain theory is used to identify four different learning styles. These styles are characterized by which region of the brain individuals prefer to use for thinking and learning. Ronald Snee (1993) summarized the relevance of these learning styles to statistical education as follows:

> Cerebral left-brain thinkers respond best when they can quantify, analyze, and theorize about things. Limbic left-brain thinkers like to see how things are put together, organize things, and practice. Cerebral right-brain thinkers learn best when they can explore ideas, discover on their own, and conceptualize what is happening. Limbic right-brain thinkers learn best when the activity is personalized and they rely on their feelings. Different learning styles require different learning methods ... Some people like to read books and listen to lectures (cerebral-left brain). Others rely heavily on doing exercises, creating summaries, and reviewing the material (limbic-left brain). Cerebral-right brain thinkers prefer visual aids, metaphors, and experiments, while the limbic-right brain thinkers like group projects, discussions, and sharing experiences.[2]

In terms of whole brain theory, you could say that Dr. Derr was assuming that Mr. Johnson was a "cerebral left-brain" learner in segment 6. However, he did not appear to gain much understanding from this presentation. Perhaps Dr. Derr made an incorrect assumption!

---

[2] From "What's missing in statistical education?" by Ronald D. Snee, p. 152. Copyright © 1995, American Statistical Association. Reprinted with permission.

A third theory of learning, the theory of neurolinguistic programming, characterizes three different learning styles. These styles are identified by the sensory modality with which individuals prefer to learn and communicate. For example, someone who relies entirely on *visual* learning needs to see information. She will read your technical reports carefully and will benefit from a well-constructed table, diagram or figure. Apparent inconsistencies or inaccuracies in a report may easily derail her understanding of your technical material. If she can't understand how something works, she is likely to read the directions. An *auditory* learner will learn more from your verbal discussion of a project than from reading a technical report. She will prefer to gain her understanding from his conversation with you, checking out her comprehension by reflecting information back to you. She probably won't be too distracted by inaccuracies in a technical report because she won't have read it that carefully. If she can't get something to work, she is more likely to ask you for help than to read the directions. A *kinesthetic* learner will prefer to have some hands-on exposure to the technical information you are providing. She will learn best by manipulating data and spreadsheets, learning the concepts by experiencing the consequences of making changes. She is not likely to read a technical report carefully. If she can't get something to work, she is more likely to keep "hacking" or experimenting until she gets it to work. Most people make use of visual, auditory, and kinesthetic modalities to some extent for learning. Some people will strongly prefer one modality to the exclusion of the other two, while others will prefer more of a balance of all three modalities.

From the point of view of neurolinguistic theory, Dr. Derr made use only of the auditory modality in segment 6. She tacitly made the assumption that Mr. Johnson was an auditory learner. Going by Mr. Johnson's apparent reaction to her spoken lecture, her assumption was not correct.

How can you make use of theories of learning to improve the way you communicate about statistics to your client? Here is what Ron Snee (1993) advised teachers to do in a classroom setting:

> ... Different learning styles are present in each group of people. People take in and process information in different ways. We want the learning process to be robust to a variety of learning and information-processing styles. Each educational experience, therefore, must include a variety of learning methods. Assigned readings and lectures get the attention of the left-brain thinkers. Others respond best to group exercises, experiments, games and metaphors. The goal is to enable each person in the group to relate to the course concept in some way.[3]

Translating this advice into a setting for statistical consulting means that you should be prepared to communicate statistical information in a variety of different ways. Then you should be

---

[3] From "What's missing in statistical education?" by Ronald D. Snee, p. 153. Copyright © 1995, American Statistical Association.

attuned enough to your client to be able to pick up clues about the ways that work best for him. Most people are not very aware of what their learning style is, so you may not learn much from a direct question. However, a common assumption among the three learning theories that we covered is that a person's preference for learning is the same as his preference for communication. This means that you can look for clues from your interaction with a client. Here is what you might look for:

(a) What has your client brought with him? If he has brought references, a detailed outline of his study, graphs, tables, some object from his study, or a computer disk, then you may want to provide something similar for him.

(b) What resources does your client use in his interaction with you? If he draws a diagram while speaking, or writes out words or phrases, then again you can try doing the same when you are talking about statistics. Does he want to make use of the computer to demonstrate something? Then try to incorporate some hands-on computer work when you are explaining statistics to him.

(c) How does your client make use of language in speaking with you? Does he use abstractions or does he provide concrete examples? Does he make use of any analogies or metaphors? Try to deliver your statistical information in a similar way.

Because statistical information is challenging to learn, statistics teachers are often advised to use more than one method in the classroom. Visual aids, written resource material, good oral presentation, oral and written directions, and opportunities for hands-on practice will enable a teacher to be effective to a broad range of students with different learning preferences. Statistical consulting is also challenging to learn. The book / video package you are making use of right now is an example of training through multiple approaches. The written text, the illustrations, the video segments, and the exercises for class discussion, individual work and teamwork have been coordinated to address the various needs of different learners. A teacher makes use of multiple approaches because of the variety of preferences that are likely to exist in a class of many students. This is also a good strategy for you as a consultant even though you might be working with just one client at a time. When you have a collection of methods available you will be able to adapt quickly to a client's apparent preferences. You can then have the maximum possible impact in the limited time available to you to explain statistics.

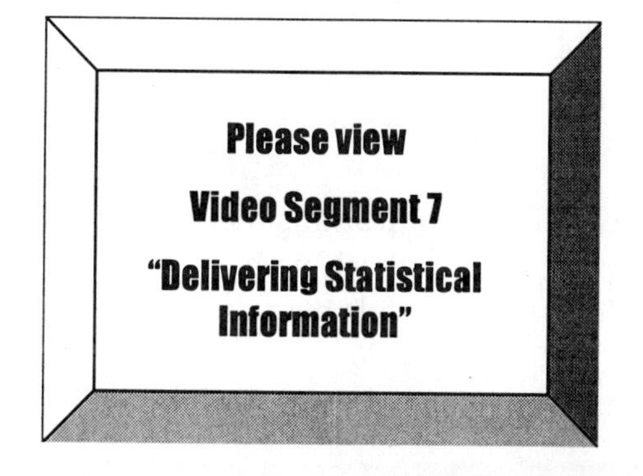

How can you deliver statistical information in a way that addresses more than one learning style? If you can, view segment 7 of the video now before continuing with this section. In segment 7, Dr. Derr once again attempts to explain her recommendations for a sampling plan to Mr. Johnson. How many differences between segment 6 and segment 7 can you identify? Do you think Mr. Johnson will implement her recommendations?

In segment 7, Dr. Derr made use of more than one method to communicate her recommendations to Mr. Johnson. In fact, before talking about her sampling plan, she showed Mr. Johnson a picture of the plan. The picture was a printout from a spreadsheet and is shown in Exhibit 7.2.

Exhibit 7.2 ***Dynamic Sampling Plan for UMS Survey in Video Segment 7***

Populations		Population size	Proportion to sample	Number going out	Expected in return
1 MAIN SAMPLE					
	DOMESTIC	38000	3.5%	1330	399
	INTERNATIONAL	2000	3.5%	70	21
		40000	3.5%	1400	420
2 SUB SAMPLE					
	INTERNATIONAL	2000	15.0%	300	90
3 SUMMARY				OUT	IN
	DOMESTIC	38000	3.5%	1330	399
	INTERNATIONAL	2000	18.5%	370	111
			TOTALS	1700	510

Mr. Johnson reacted immediately and positively to this picture, which was a clue that he preferred visual displays of information. Because of his reaction, Dr. Derr continued to make use of the picture to explain the sampling plan. They talked to each other about the target population and the sampling plan while pointing to various parts of the picture. She also asked him if he were familiar with spreadsheets from his work with budgets: "I'm laying this out as a spreadsheet and I'm wondering if you have used spreadsheets before." She used an analogy, "think of this as a sampling spreadsheet … and it's the same idea as a budget spreadsheet" to connect her statistical work with something more familiar to him. She then provided him with a dynamic demonstration of the spreadsheet. When she entered a different percentage to be sampled in the shaded boxes, the spreadsheet outputs changed. She demonstrated this on the computer while he watched and then gave him the diskette for him to work on after the meeting. This gave Mr. Johnson an opportunity to develop his understanding further through experimentation. He received this diskette with evident delight.

Exercise 7.1

What is your preferred learning style? Consider each theory of learning that has been described in this section. How would you characterize yourself with respect to the styles described by each theory? More than one style may suit you.

Exercise 7.2

In an article about a laboratory-based method for teaching statistics to engineers (Barton and Nowack 1998), one of the discussants had this to say about how engineers learn statistics:

Bluntly speaking, our traditional statistics courses lack engineering relevance and are too boring. ... Many educators continue to insist that theory must precede examples and applications. However, engineers and other people with a practical bend prefer to see examples and motivation first. Thus, Barton and Nowack's hands-on lab approach should be enthusiastically promoted. ... Engineering students are accustomed to lab work and will, with this approach, see statistics at work rather than as abstract classroom theories. Thus, the lab approach can go a long way to solve the motivational problems we might have had in the past.[4]

Based on this comment, how would you characterize the learning preferences of engineers? What do you think would be the best ways to communicate about statistics to a client who is an engineer?

7.5 Talk about Statistics

A big difference between segments 6 and 7 was the amount of interaction between Dr. Derr and Mr. Johnson. In segment 6, Dr. Derr lectured at Mr. Johnson without giving him much opportunity to contribute or provide feedback. In segment 7, Dr. Derr collaborated with Mr. Johnson. She gave him the opportunity to state his impressions and provide feedback. In this way Dr. Derr made sure that Mr. Johnson followed the statistical discussion and that her recommendations truly addressed his concerns. She paused between comments to permit Mr. Johnson to provide some feedback, and she made sure that he was comfortable about expressing his lack of understanding. She also appeared to be alert to any clues that Mr. Johnson had not fully comprehended what she said. You may also have noticed that segment 7 is nearly twice as long as segment 6, even though the statistical content of the two segments is about the same. It does take more time to discuss statistics in an active learning mode compared a passive learning mode. However, you will find that this additional time is an investment in comprehension, feedback and acceptance that will yield great benefits.

Some people are very reluctant to say they don't understand something. A client may interpret his own lack of understanding as a loss of face for himself or even as a loss of face for you. He may be so apprehensive about statistics that he doesn't really listen to what you have to say. This is why you should be especially cautious about interpreting a nodding head and comments such as "yes", or "I see", as a sign that your client really understands what you are saying. This is

---

[4] From "A one-semester, laboratory-based, quality oriented statistics curriculum for engineering students""by R.R. Barton and C.A. Nowack, p. 238. Copyright © 1998, American Statistical Association. Reprinted with permission.

where your earlier preparation with this client should start to pay off. When you took the time to establish rapport with your client, you helped him to feel comfortable about expressing his lack of understanding. When you paraphrased what he told you, you provided him with a model of paraphrasing as a way of clarifying what someone else has said. The best opportunity you have to assess whether your client understands your comments comes when he begins to work with the information. This can start with a paraphrase. In segment 7, Mr. Johnson pointed to cells in the spreadsheet and paraphrased what he and Dr. Derr had been discussing about the sampling plan: "So this is over the main sample and then we go to the sub-sample here and that's a little bit higher number … but if you took this number up here it would be far too costly." You can facilitate a discussion like this with a display of some kind, whether it is a table or figure, a computer program, or some other model or prop. In segment 7, Mr. Johnson made frequent reference to the printout of the spreadsheet. In fact, he set the pace of the discussion with his observations, such as:

> *"And the breakdown here is with the internationals..."*
>
> *"So 3.5% we're going to take and it's like forty thousand..."*
>
> *"And this is the expected return..."*
>
> *"This is the sub-sample here?"*
>
> *"And that's only twenty one coming back..."*

Dr. Derr used each one of these observations to build on her discussion of the sampling scheme.

Now that you have established a framework for the discussion, let's consider how to interpret the language of statistics for the client. As you probably noticed in the negative version of the video, Dr. Derr made no effort to clarify the statistical terminology she used in her lecture. She used terms such as "expectation," "categorical," "dichotomous" and "95% lower bound" without any further explanation. Dialog 1 is a transcription from segment 6 that contains her verbal explanation of sampling error and its relation to the decision about sample size. This jargon-filled delivery had the effect of excluding Mr. Johnson from the discussion. If you were to talk about statistics in such a way to a real-life client, he may not be as acquiescent as Mr. Johnson was. Your client might get angry with you! He certainly would be justified in telling his colleagues that you are impossible to understand.

What is the best way to discuss statistical terms and concepts? Statistical terms are difficult for two reasons: (1) your client may not be familiar with them; and (2) they are abstractions. Terms like "confidence interval", "p-value", and "dichotomous" may carry little meaning to your client. He may question the relevance of these terms to the goals of his study. In order to be clear, you must first translate statistical terms into words that the client can understand and then provide some concrete examples that show their relevance to your client's study. This will encourage your client to take the time and energy required to understand the statistical terms. Finally, be prepared to make a multi-media approach to your explanation. Consider how you might make

Dialog 1 ***Transcription of a portion of video segment 6***

> When we talked about what was adequate about what was coming in … we said maybe 100 for the internationals coming back in and maybe 400 for the main group coming in. But we didn't really base that on anything, that was just … a comfort level. … Most of your questions that you showed me on the questionnaire are categorical, and I simplified that by saying that they were dichotomous … so I said, what would be the 95% confidence interval of about 50% of the respondents saying "yes" to one of your questions or … some dichotomous response. With 100 responses you have a 95% confidence interval that goes from 40% to 60% if your true population is 50%. With 400 that interval narrows to 45% to 55%. So you see what happens with a bigger sample size, you get more precision in your estimates. Now it's going to be up to you and your committee to decide if that level of precision is adequate.

use of diagrams, tables, references, and the computer to augment your discussion. This will enable you to adapt quickly to whichever approach appears to be the most fruitful for your client. In segment 7, Mr. Johnson's interest in the spreadsheet program was a clue that he preferred to be a "hands-on" learner. Taking advantage of this clue, Dr. Derr used the dynamic medium of the computer to help explain the concept of sampling error and how this concept related to the decision about sample size. A transcript of this discussion is in Dialog 2, and the computer display is in Exhibit 7.3. In Dialog 2, Dr. Derr provided a concrete and relevant example for the hypothetical response of 50% plus or minus a margin of error. The computer demonstration also helped convey the relationship of sample size to precision. Since Mr. Johnson had the spreadsheet file, he could continue to develop his insight about precision and sampling further after the meeting. Some of your clients who learn well by reading may also appreciate having a written summary from you and user-friendly references about the statistical topics you have discussed.

Dialog 2 ***Transcription of a portion of video segment 7***

Dr. Derr	The last thing here ... I just wanted to show ... We have 400 coming back from the main sample and 100 coming back in from the internationals. I want to show you what that does with the precision of their responses. *(points to bar chart)*... What you have here is a bar chart showing the margin of error for each of the groups, the domestics and the internationals. ... Because you have a bigger sample with the domestics you have a smaller margin or error. Now I don't know if that term is familiar ... to you.
Mr. Johnson	... Margin of error means the possibility of being wrong? Error sounds to me ... *(laughs)*
Dr. Derr	*(Laughs)* ... Error sounds like you're wrong. Actually ... it represents what the total population might have said had you censused them all. Here you're sampling a small percentage of the population ... For example, let's say 50% of your sample said they were satisfied with the location of your medical services. Then had you censused the whole population then somewhere between 45 and 55 [percent] would be happy with the location.
Mr. Johnson	I understand.
Dr. Derr	So this whole thing actually is driven by these inputs ... As you put in the different percentages, these margins of error will get smaller or larger depending on how big the sample size is.
Mr. Johnson	This is also on my disk?
Dr. Derr	That is on your disk and I'll just show you here on the spreadsheet *(scrolls down)* ... It's these bars that will get bigger or smaller depending on how big your sample is. ... So it's something you can take to your committee, to see how much precision do they require to make decisions ... about improving some aspects of their medical services.

Exhibit 7.3 ***Dynamic Error Bars from UMS Sampling Plan (Video Segment 7)***

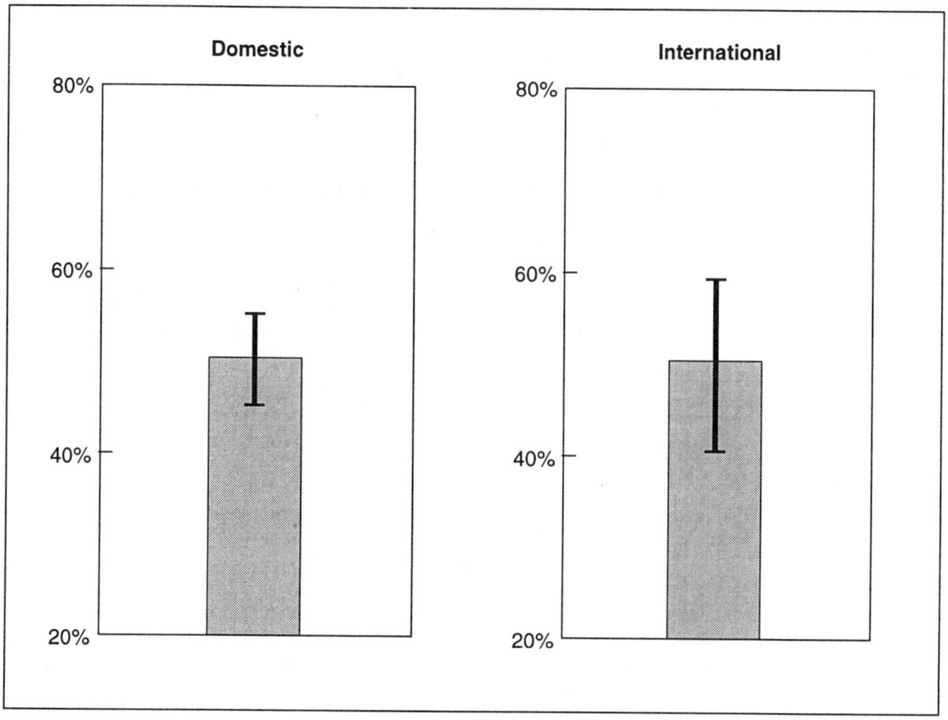

Domestic

80%
60%
40%
20%

International

80%
60%
40%
20%

Exercise 7.3

A statistical consultant is often called upon to provide an explanation of a statistical concept in his conversation with a client, with very little notice. This exercise will give you some practice in doing just that!

You may work in small groups of two or three. When you are ready, select a statistical term or concept at random. You may select from the list in this exercise or from some other source. Give yourself a minute to prepare, then deliver a verbal explanation of this concept to a member of your group who is serving as the hypothetical client. Do this by making use of the auditory modality only (no pictures, drawings, graphs, or diagrams!). Once you are finished, have the other members of your group critique your delivery. Each member of the group should conduct this exercise in the role of consultant.

Here are some statistical terms and concepts to get started. Feel free to add others.

- The concept of covariates in an analysis of variance model.
- The meaning of bias in a survey.

- Correlation.
- The problem with multicollinearity in a multiple regression model.
- A dichotomous response.
- Regression to the mean.
- The effect of measurement error in the independent predictors of a linear regression model.
- The use and interpretation of the log transform of the dependent variable in a linear model.
- An odds ratio.
- The difference between a paired t-test and an independent t-test.
- Autocorrelation in samples taken over time.

Exercise 7.4

Exercise 7.3 was probably pretty frustrating! It is difficult to be clear, using just words alone. Now lift those restrictions. Choose several statistical concepts, either from the list in Exercise 7.3 or others. For each one, describe how you would develop a multi-media approach to explaining these terms to a client. Try to include as many modalities as you can from the following list, and you may certainly use other means as well:

- A figure
- A table
- A computer demonstration
- User-friendly references
- A written summary
- Concrete examples from the client's field of application (make one up, borrow from your experience, or refer to one of the examples in this book).

7.6 Present Statistics to an Audience

What should you do if your clients invite you to make a more formal presentation to them? This is very likely to happen! Gerry Hahn and Roger Hoerl (1998) made this comment on communication in business and industry:

> Detailed written reports are being downplayed in favor of concise management-oriented presentations - often presented remotely via video or teleconferences. Statisticians need to adapt to these new forms of communication[5].

Your clients may put you on the agenda along with other presenters, as a way of gathering information and opinions about an issue or decision. A presentation is a good opportunity for you to make the best case for your statistical ideas. The group's attention will be focused on you and what you have to say and show to them. Fortunately, there are many excellent resources you can use to improve your presentation skills, and some of these are listed at the end of this chapter. In this section we will cover some suggestions for ways that you can make the most of this opportunity to communicate your ideas to your clients.

In Example 7.5, Elaine Walker, the statistician for the FDA, was called upon to give a presentation to a review committee. This committee had been charged with the task of evaluating a veterinary device that was used in clinical practice. The device had been the subject of numerous complaints from users, and it was the committee's responsibility to decide how to respond to these complaints. Elaine had worked with Vincent Cantrell, the analytical chemist who was in charge of the laboratory assessment of the device. Together they developed a protocol, and Elaine had analyzed the data from the protocol. Her presentation was one of several on the meeting's agenda. We will use this example to illustrate the following suggestions about how to present statistical information.

1. Understand the needs and expectations of your audience. Make sure you know who will be in your audience, and what they expect you to talk about. This presentation is not like a statistics lecture in a classroom, nor is it similar to a presentation of your statistical research to your peers. Instead, you may be speaking to managers, administrators, investigators, and others who need to make use of your information in order to make decisions. They want to know what your core message is and how it relates to their concerns. They will not be evaluating the statistical accuracy of your work on the basis of your talk.

The data analysis that Elaine used to evaluate the measurement error of this device was lengthy and complicated. Not only did the study generate a large amount of data, but the protocol was designed around several factors that had to be modeled in the analysis. However, she knew that this review committee did not want to hear about the lengths that she had gone through to

---

[5] From "Key challenges for statisticians in business and industry" by Gerry Hahn and Roger Hoerl, p. 196. Copyright © 1998, American Statistical Association. Reprinted with permission.

analyze the data. The committee was made up of administrators, many of them with a scientific or clinical background, who needed to know about the measurement error of this device. This was one of several inputs about the device that would influence their decision about how to address the complaints about it. Therefore, Elaine knew she had to provide summaries that would enable the committee to understand the extent and nature of the measurement error of the device.

2. Make sure your main points address the audience's objectives. If you have spent a lot of time, talent and energy figuring out the details of a study design or analyzing data, it is only natural to want to talk about these details as you might to a statistical audience. After all, most of us chose the profession of statistics because we enjoy this part of the process! However, the non-statisticians and decision-makers in the audience are most interested in having your statistical input to their problems. Your presentation should be organized around their objectives. Be clear and direct, and follow the priorities of your clients.

The first slide of Elaine's presentation (Exhibit 7.4) listed the objectives that connected the study to the committee's concerns. A colorful cartoon on the slide added humor and served as the analogy that Elaine used to explain the basic concepts of measurement error to the audience. At the end of her talk, Elaine added a short summary of her main statistical findings on to this same slide (Exhibit 7.5). By beginning with the objectives and ending with a brief summary of the statistical findings associated with each objective, Elaine made sure that the committee was aware of the main findings of the study and could incorporate these findings in their decision about the device.

3. Know what your time limit is and stay within it. This is good advice for all speakers. If there are several presenters on the agenda, it is a courtesy not to infringe on their time. A key to working within your time limit is to rehearse. If you have more material than your time limit provides, don't speed up your speech! Instead, prune your material. A few good points well presented will stay with your audience much longer than a galloping speech with too many visuals. If there is much more that you would like your audience to know about, write it up and hand it out.

4. Use excellent visuals. The quality of the visual aids you use is likely to influence what your audience thinks about your statistical message. They are not in a position to evaluate your technical skills, but they can assess the quality of your presentation. They are also likely to be accustomed to high quality visuals used in other domains of their work environment. This is not the time to scribble formulas on a transparency with a felt tip pen, or to copy from a reference directly onto a slide (no one can read this!). The technology for producing visuals today is user-friendly enough so that you can easily build visuals that enhance your message with color and style, and even with motion.

Elaine wanted to depict the study design to the committee. She used a diagram (Exhibit 7.6) showing one week of the multi-week study. She used color and shading to guide the committee

Exhibit 7.4 *Introductory Slide of Elaine's Presentation*

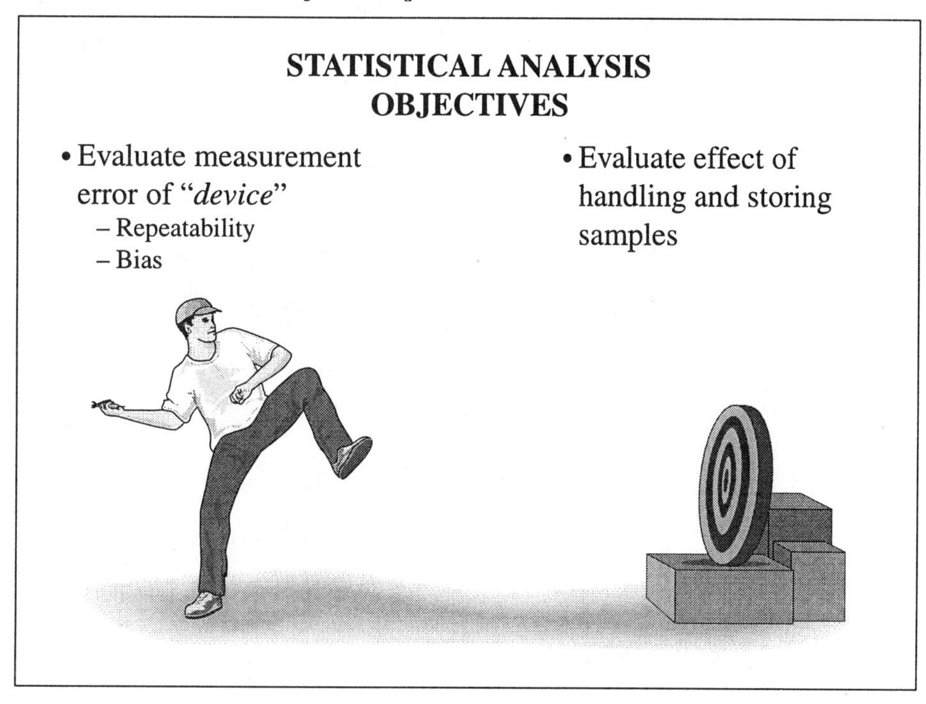

Exhibit 7.5 *Conclusion Slide of Elaine's Presentation*

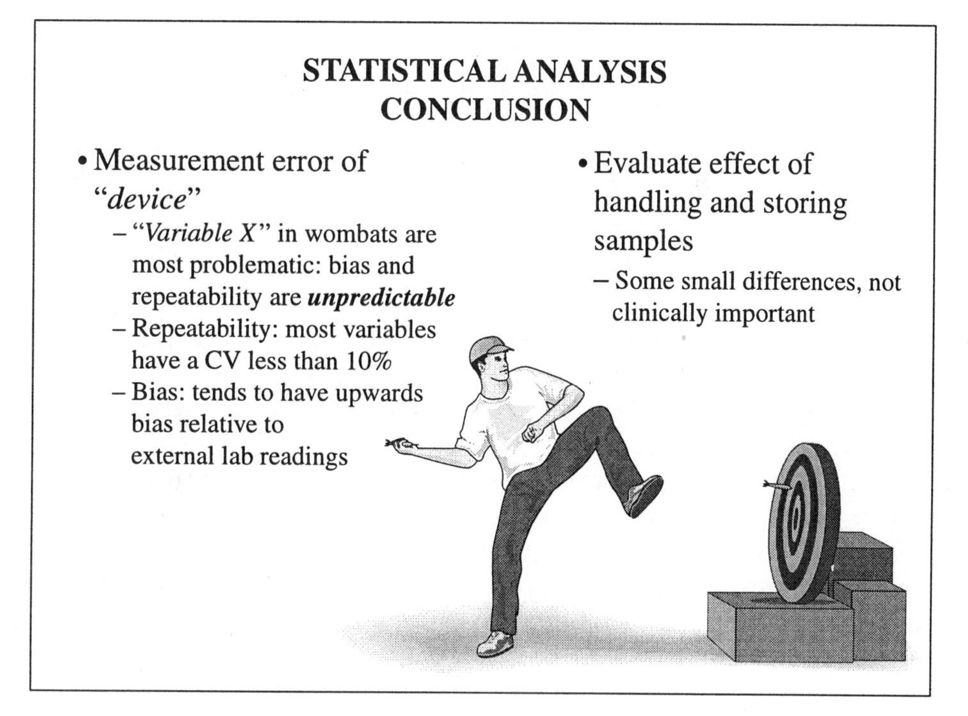

through the collection of data from one day on one species. This permitted her to expand the design with her words to encompass the number of days in the study, the number of animals per species, the number of replicates per animal, and the other factors around which the study was structured.

Exhibit 7.6 ***Elaine's Presentation of the Study Design***

Week 1		DAY 1		DAY 2		...	DAY 4	
Protocol	**Operator**	**Species**	**Sample**	**Species**	**Sample**		**Species**	**Sample**
Imme-	OP #1	Quoll 1	Rep 1	Quoll 3	Rep 1	...	Quoll 7	Rep 1
diate			Rep 2		Rep 2			Rep 2
		Quoll 2	Rep 1	Quoll 4	Rep 1		Quoll 8	Rep 1
			Rep 2		Rep 2			Rep 2
		Wombat 1	Rep 1	Wombat 3	Rep 1		Wombat 7	Rep 1
			Rep 2		Rep 2			Rep 2
		Wombat 2	Rep 1	Wombat 4	Rep 1		Wombat 8	Rep 1
			Rep 2		Rep 2			Rep 2
	OP #2	Quoll 1	Rep 3	Quoll 3	Rep 3		Quoll 7	Rep 3
			Rep 4		Rep 4			Rep 4
		Quoll 2	Rep 3	Quoll 4	Rep 3		Quoll 8	Rep 3
			Rep 4		Rep 4			Rep 4
		Wombat 1	Rep 3	Wombat 3	Rep 3		Wombat 7	Rep 3
			Rep 4		Rep 4			Rep 4
		Wombat 2	Rep 3	Wombat 4	Rep 3		Wombat 8	Rep 3
			Rep 4		Rep 4			Rep 4
Delayed	OP #1	Quoll 1	Rep 5	Quoll 3	Rep 5		Quoll 7	Rep 5
			Rep 6		Rep 6			Rep 6
		Wombat 1	Rep 5	Wombat 3	Rep 5		Wombat 7	Rep 5
			Rep 6		Rep 6			Rep 6
		Quoll 2	Rep 5	Quoll 4	Rep 5		Quoll 8	Rep 5
			Rep 6		Rep 6			Rep 6
		Wombat 2	Rep 5	Wombat 4	Rep 5		Wombat 8	Rep 5
			Rep 6		Rep 6			Rep 6
	OP #2	Quoll 1	Rep 7	Quoll 3	Rep 7		Quoll 7	Rep 7
			Rep 8		Rep 8			Rep 8
		Wombat 1	Rep 7	Wombat 3	Rep 7		Wombat 7	Rep 7
			Rep 8		Rep 8			Rep 8
		Quoll 2	Rep 7	Quoll 4	Rep 7		Quoll 8	Rep 7
			Rep 8		Rep 8			Rep 8
		Wombat 2	Rep 7	Wombat 4	Rep 7		Wombat 8	Rep 7
			Rep 8		Rep 8			Rep 8

5. **Seek input from your clients during your preparation.** Your clients can give you the best suggestions about how to present your ideas to them. If you can't prepare directly with a few members of your audience, then find some people from similar backgrounds who are willing to give you some feedback. It is better to find out before your talk that nobody can interpret your visual aids or understand what you are saying than to be staring at a wall of blank faces. Just as with an informal conversation with a client, your best indicator of whether or not your audience understands what you are saying is if they are able to work with the material you are presenting. Someone with a similar background as those in your audience can help you improve the clarity of your presentation.

The most important part of Elaine's talk was to present the results that had to do with the measurement error of the device. She wanted this part of the talk to have as much impact as possible. She debated about how to display the results of her analysis: as figures? In a table?

She also wondered which summary statistics would be meaningful to the audience. She enlisted the assistance of two co-workers, who had a clinical perspective that resembled the backgrounds of her audience. She showed them drafts of figures and tables, and asked them for their input. The summary table she thought was very clear meant practically nothing to them. One co-worker said, "I'd just like to see a picture of the raw data for each animal, so I could see for myself how varied the replicates are." Once she produced such a picture both co-workers began assessing and interpreting the data from their perspectives as practitioners. In this process, Elaine learned a great deal about how this device was used in veterinary practice, and what aspects of measurement error would cause the biggest problem for a practitioner. For her presentation, Elaine produced several figures that showed the variability of the replicates for each animal. Exhibit 7.7 depicts this image for one species; the actual presentation slide had four such graphs, one for each species, with a common scale for improved comparison among species. She also produced a table with the summary statistics about bias and repeatability. In her talk she was able to refer the audience to the summary statistics in the table after they had developed their impressions of the data from the figures. This assisted the audience in their interpretation of the summary statistics.

Exhibit 7.7 ***Figure from Elaine's Talk Demonstrating Variability among Replicates***

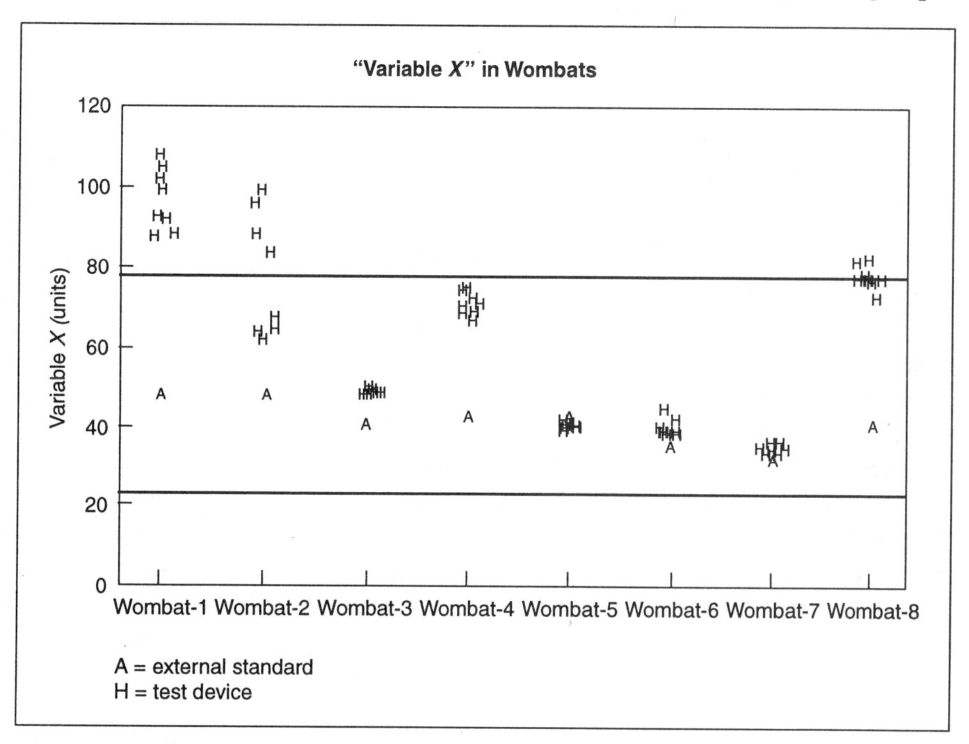

6. Provide a handout that reinforces your talk. A handout with copies of your visuals and other information will enable members of your audience to see the visuals up close, and also look at other supplementary material that will improve their understanding of what you have to say. They can then refer to the handout after your talk to recall your main points.

Elaine included figures for all of the variables that were being assessed, in addition to a comprehensive table in the technical report. For her talk, she only concentrated on a few representative variables. Both the technical report and copies of the slides were circulated to the committee at the start of the meeting.

Exercise 7.5

Suppose that part or all of your audience was attending your presentation by: a) videoconference b) telephone conference call. How would you adapt your presentation accordingly? You can refer to Chapter 4 for suggestions.

Exercise 7.6

If possible, give a brief presentation based on an actual consulting problem that you are working on. Obtain feedback about your presentation, including comments about your visuals.

7.7 Write about Statistics

Although presentations may be replacing written reports in many consulting situations, there will still be occasions when you will need to communicate in writing. As you can tell from the examples in this chapter and elsewhere in this book, the length and formality of these written communications is highly varied. Email is written communication and is usually brief, more in the style of a written memo. Executive summaries are also brief and directed towards a non-statistical audience. A formal technical report may have different sections that are written to different audiences. You will probably also be using written records to document your agreements with your client and the process of your statistical work with the client's data. Add to this letters to clients and summaries of your meetings and it seems as if there will be plenty of opportunity for you to practice your skills in writing! Fortunately there are many resources to help you develop your writing skills, and some of these are listed at the end of this chapter. In this section we will cover some suggestions for improving the clarity of your writing when it is directed to your clients.

1. Identify the audience. This suggestion is beginning to sound very familiar! Probably the biggest step you can take towards clarity in any communication is to be aware of who your audience is. As statisticians we sometimes forget that we are not writing to other statisticians in these communications. Instead, consider the types of people who will be reading your document and address them accordingly.

In Example 7.4, James High faxed a table to the evaluation team that summarized the costs and benefits of the several options they were considering. This table would serve as a basis for their telephone conference call. In the WEST study (Example 7.1), Jean prepared a technical report that had two audiences: (1) the program director of the study, who needed to be able to interpret the results of Jean's analysis and (2) statisticians who may at some future time want to understand the details of the data acquisition and analysis. She addressed these two audiences successfully by separating the technical details of the study into a technical appendix. In the UMS survey that provides the story line for our video (Example 7.3), Dr. Derr prepared a notebook for Mr. Johnson to examine that contained preliminary summary tables. Later, she prepared a short report for the committee and their stakeholders. She also wrote a technical report that contained the details of the sampling plan, the database, and the response statistics. This technical report was addressed to technical specialists in survey design and analysis.

2. Assume that your client is very busy and write accordingly. Of course we'd like to think that what we have to say is so important that clients will drop everything and read every word that we have written. However, most people have multiple demands on their time. They would like to understand the core message of a document without laboring to find it. Therefore, make sure that the highest priority of your document is to address your client's concerns. An executive summary at the front of the report is a good place for you to summarize the key findings or recommendations that address the most important issues of the project. This executive summary is directed to the client and is written in clear language that he can understand. You can then document the statistical details of your work in a technical appendix. You can write this appendix to a statistical audience. An example of an executive summary from the report that Dr. Derr wrote about the UMS survey is shown in Exhibit 7.8.

As you will discover in segment 8, Dr. Derr did not consider her audience carefully enough when she delivered her notebook to Mr. Johnson. The 100-page notebook filled with tables overwhelmed him. He declared that he did not have time to sort through the materials in this notebook. This interaction threatened to derail the good relations that the two had previously established. Although strictly speaking he had asked for summaries of each question and a separate question-by-question summary cross-tabulated by origin of the respondent (domestic or international), he probably did not consider what this request would produce. Although we will not be covering segment 8 in detail until Chapter 8, you may wish to view this segment now to view Mr. Johnson's highly negative reaction to this notebook!

Exhibit 7.8 ***Executive Summary of the UMS Survey***

University Medical Services Survey

EXECUTIVE SUMMARY

The University Medical Services conducted a survey in order to find out how students felt about the health services available on campus and what attributes of health care were important to them. A questionnaire was mailed late fall semester to a randomly selected sample of 2,000 students. There was a response rate of 30%.

The majority of respondents felt that all seven health care attributes listed in the questionnaire were very important. There were no substantial differences in the responses from the international students and those from the domestic students in the survey overall, although some international students (2%) indicated concerns about language barriers.

Based on the chart below, it appears that there is room for improvement in the top three attributes viewed as very important by the most respondents: *Quality of Health Care Provided, Professionalism of the Staff*, and *Sensitivity of the Clinical Staff. Waiting Time*, while not viewed as important as other attributes, still received lower satisfaction ratings.

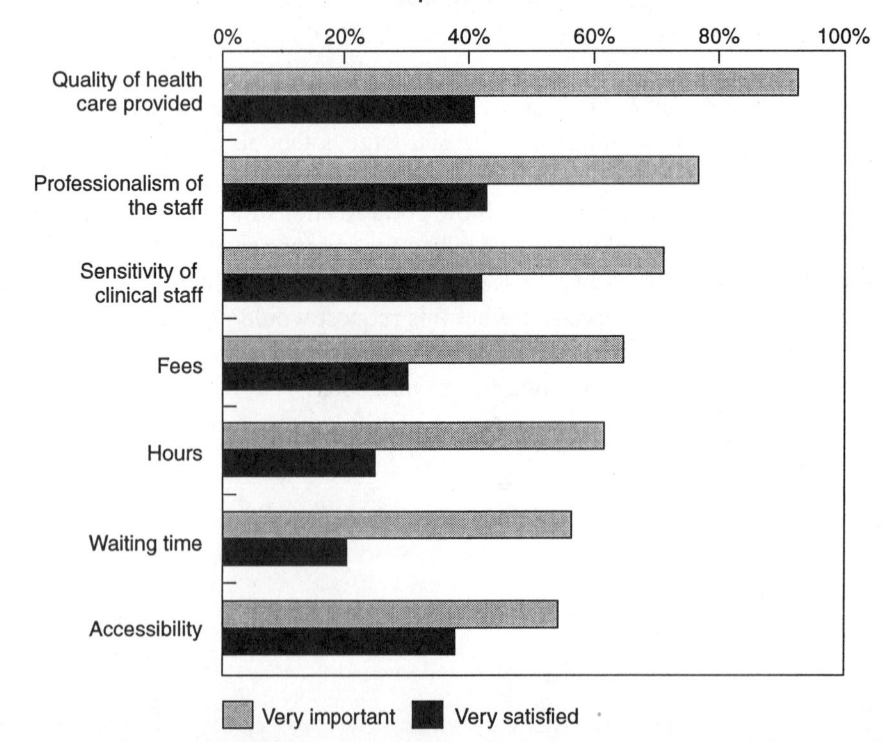

3. Make good use of tables and graphs. Figures and tables will help you communicate your message. When well prepared, they can be especially helpful to a non-statistical audience.

Exhibit 7.9 includes just two of the tables that were in the 100-page notebook that Dr. Derr gave to Mr. Johnson. These tables summarize some of the demographic features of the respondents. These are well-formatted tables from Dr. Derr's point of view. She wanted Mr. Johnson to examine them and help her decide how to calculate the final response percentages. However, there was still too much detail for Mr. Johnson to work through on his own. He probably also was overwhelmed by the sheer number of them.

Exhibit 7.9 ***Demographics of Respondents to the UMS Survey (Draft Version)***

Question 12: Gender

		Frequency	Percent	Valid Percent	Cumulative Percent
Valid	Male	238	39.7	41.3	41.3
	Female	338	56.3	58.7	100.0
	Total	576	96.0	100.0	
Missing	No Answer	19	4.0		
	Total	19	4.0		
Total		600	100.0		

Question 18: Citizenship

		Frequency	Percent	Valid Percent	Cumulative Percent
Valid	Domestic	470	78.3	81.1	81.1
	International	110	18.3	18.9	100.0
	Total	580	96.7	100.0	
Missing	No Answer	20	3.3		
	Total	20	3.3		
Total		600	100.0		

In segment 9 of the video, Dr. Derr presents Mr. Johnson with some improved summaries of his survey. One of these is a table with a more compact summary of the characteristics of the respondents (Exhibit 7.10). This table is much easier for a non-statistician to scan and interpret.

Dr. Derr also showed Mr. Johnson two graphs that summarized key parts of his survey. A pie chart depicted the distribution of the respondents from different parts of the world. A bar chart showed the way that respondents ranked the importance of seven different attributes of health care (Exhibit 7.11). The fact that the bars are ordered by their rank of importance helped Mr. Johnson interpret the results. Although we will not cover segment 9 of the video in detail until

Chapter 8, you may wish to view this segment now. You will notice that Mr. Johnson reacted very positively to both of these graphs. He was able to make some interpretations of his results immediately upon seeing the bar chart that ranked the importance of health care attributes.

Exhibit 7.10 ***Demographics of respondents to the UMS Survey (Final Version)***

Characteristic	N[1]		
Gender	576	41%	Male
		59%	Female
Degree status	571	75%	Undergraduates
		23%	Graduates
		2%	Non-degree
Residence	574	48%	On campus
		52%	Off campus
Citizenship	580	81%	Domestic
		19%	International

[1]Number of valid responses. Not all respondents answered all questions.

Exhibit 7.11 ***Bar Chart from UMS Survey***

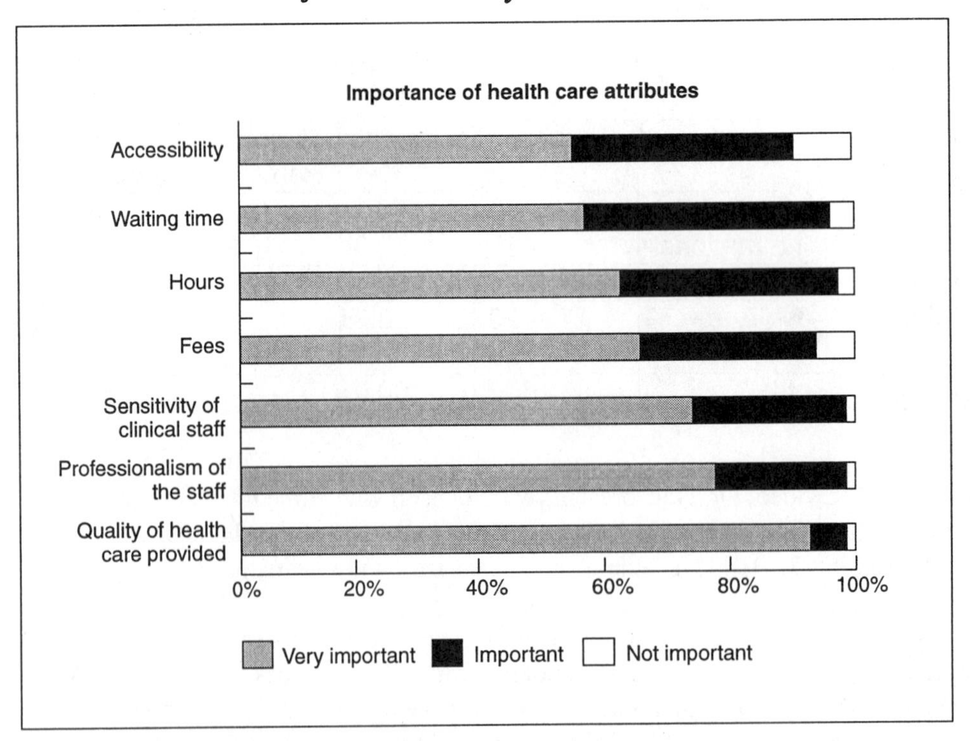

4. ***Provide your readers with an easy way to navigate through the document.*** If a document has more than one page, number them (consecutively!). Number or otherwise uniquely identify all figures, tables and anything else you may need to refer to. If the document is long, divide it into sections and include a table of contents. This may all seem very obvious. However, I have attended many meetings where the participants were leafing, puzzled, through the (un-numbered) pages of a report produced by a statistician, trying to find the (un-numbered) figure or table that the statistician was talking about. It should be no surprise to you that this practice does not enhance the credibility of the statistician's message. Think how much easier it is to be able to say "Please turn to Figure 2 which is on page 5 of your report." Then you will all literally be on the same page!

5. ***Seek input from your clients.*** It is always very helpful to have someone read a draft of your document and give you input on the clarity of your statistical message as well as other elements of your writing. Someone who can critique your document from a client's perspective can also help you identify how to improve your summaries, tables and figures so that your statistical message comes through clearly to the intended audience.

Exercise 7.7

Put yourself in the place of James High, the statistician with the telecommunications company in Example 7.4. Suppose that your team is considering three different options for finding out how consumers react to the new telecommunications product. These options are: (a) focus groups interviews; (b) a telephone survey; (c) a mail-in survey. Produce a table or exhibit that summarizes the advantages and disadvantages of each method. This should be something that you intend to fax to the other members of the project team prior to your telephone conference call. Recall that the team includes statisticians, a survey research specialist and members of the marketing group. Feel free to add your own details about the product and the consulting situation.

Exercise 7.8

Put yourself in the place of the statistician in the Chocolate Bar Study, which was described in Example 5.2. At the planning stages of this study, the investigators wanted to decide whether to conduct the study in a two-period crossover design or as a parallel arm design. One of several considerations is the statistical power of the comparison between the test diet (Chocolate Diet) and the reference diet (Pretzel Diet). The investigators would like to make sure that the study has sufficient power to detect a difference of at least 10 mg/dl of total cholesterol (TC) between the mean for the Chocolate Diet and the mean of the Pretzel Diet. From previous studies you have an estimate of the population mean TC of 160 mg/dl for people in this demographic group, the between-subject standard deviation of 20 mg/dl, and the within-subject standard deviation of 12 mg/dl. Develop a figure depicting the statistical power of each proposed design. Write a summary, directed to the investigators, of the advantages and disadvantages of each design.

Exercise 7.9

Put yourself in the place of Jean Reina, the statistician in the WEST study in Example 7.1. Develop figures from the results shown in Tables 1 and 2. Include the figures in an executive summary of the results. Write this summary to the program administrators who want to know about the effectiveness of mentoring programs similar to WEST.

Here is some more information about the WEST study (You can also review the description in Example 5.4): The goal of the WEST program was to promote the retention of undergraduate women in non-traditional science and engineering majors. The "WEST" students were first-year undergraduate women who participated in a special mentoring program during their second and third semesters. For this observational study, Jean searched the college's data warehouse and obtained a match for each WEST student with a similar male and a similar female (non-WEST) student. She then tracked each student's progress through the undergraduate program for as many semesters as was possible, noting any change of major.

Table 1. Number of students in each group who changed majors at least once.

Group	Change in Major	No Change in Major	Total
WEST students	20	52	72
Control (Female)	28	44	72
Control (Male)	27	45	72
Total	75	141	216

Table 2. Number of students in each group moved out of a science or engineering (SEM) major.

Group	Change in Major	No Change in Major	Total
WEST students	8	64	72
Control (Female)	20	52	72
Control (Male)	18	54	72
Total	46	170	216

Exercise 7.10. ***Will Computers Replace Statistical Consultants?***

Some statisticians perceive the development of user-friendly statistical software and interactive decision-making software as threats to the livelihood of statistical consultants. What do you think? What are arguments for and against this position? Find out what other experienced statisticians and clients think.

7.8 Suggestions for Group Discussion

1. Exercises 7.3 and 7.4 can be implemented in a variety of ways. Do the verbal descriptions of statistical concepts for Exercise 7.3 in class, with only a brief amount of time for preparation. This will simulate the need for ready explanations within a consulting discussion. Set a time limit for these explanations; five minutes should be more than enough. You can record the participants' delivery of their verbal statistical explanations. Students can then prepare a written transcript of what they said from the recording. Exercise 7.4 can then be done outside of class, so that the students have more time to prepare. You can adapt this exercise so that the statistical topics are relevant to the consulting projects that the class is involved in. You can also involve clients in this exercise as follows: Ask clients to contribute statistical terms and concepts that are important to their work. Use these concepts for Exercises 7.3 and 7.4. In class, with the students' consent, have the clients listen to the recorded explanations and provide constructive feedback. Then ask the clients to comment on the multi-media approaches proposed by the students for Exercise 7.4.

2. To increase the benefit of Exercise 7.6, you can do some or all of the following:
   - Have the students prepare a presentation based on their actual consulting projects.
   - Ask clients to review the draft materials of the presentation and give feedback to the student.
   - Have the students present to an audience that consists of statisticians and clients.
   - Ask the audience to fill out a feedback form. Make sure the students see the form in advance so that they know how they will be evaluated. You can even enlist their help in developing the format for evaluation.
   - Videotape the presentations and provide feedback to the students.

3. You can also have the students write material such as executive summaries and recommendations for work that they are actually doing with clients.

4. Exercise 7.10 can be adapted in several ways. It could be a topic of a writing assignment, or of a debate or a panel discussion. You can invite others to participate in this discussion: clients who make extensive use of statistical software, experienced statisticians and software developers. Choosing people with opinions on all sides of the issues will enliven the debate!

7.9 Resources

Barton, R.R. and Nowack, C.A. (1998), "A one-semester, laboratory-based, quality-oriented statistics curriculum for engineering students," *The American Statistician,* 52, 233-43.

Becker, R.A., and Keller-McNulty, S. (1996), "Presentation myths," *The American Statistician*, 50, 112-115.

Blyler, N.R. and Thralls, C. (1993), *Professional Communication: The Social Perspective,* Newbury Park, CA: SAGE Publications.

Cobb, G. (1991), "Teaching statistics: More data, less lecturing," *AmStat News,* 182, 1-4.

Freeman, D.H. Jr., Gonzalez, M.E., Hoaglin, D.C., and Kilss, G.A. (1983), "Presenting statistical papers," *The American Statistician,* 37, 106-110.

Hahn, G. and Hoerl, R. (1998), "Key challenges for statisticians in business and industry," *Technometrics,* 40, 195-200.

Hamilton, C. and Parker, C. (1993), *Communicating for Results: A Guide for Business and the Professions,* Belmont, CA: Wadsworth, Inc.

Huckin, T.N. and Olsen, L.Z. (1991), *Technical Writing and Professional Communication for Nonnative Speakers of English,* NY: McGraw-Hill, Inc.

Kolb, D.A. (1984), *Experiential Learning--Experience as the Source of Learning and Development,* Englewood Cliffs, NJ: Prentice-Hall.

Lazear, D.G. (1990), *Seven Ways of Knowing: Teaching for Multiple Intelligences*, Australia: Skylight Publishing, Inc.

Maindonald, J.H. (1992), "Statistical design, analysis, and presentation issues," *New Zealand Journal of Agricultural Research*, 35, 121-141.

McKeachie, W.J. (1995), "Learning styles can become learning strategies," *The National Teaching and Learning Forum*, 4, 1-3.

McPhee, D. (1996), *Limitless Learning: Making Powerful Learning an Everyday Event,* Tucson, AZ: Zephyr Press.

Nemeth, M.A. (1998), "Discussion," *Technometrics,* 40, 206-207.

Snee, R.D. (1993), "What's missing in statistical education?" *The American Statistician,* 47, 149-154.

Tufte, E. (1983), *The Visual Display of Quantitative Information,* Chesire, Connecticut: Graphics Press.

8

DEALING WITH DIFFICULT SITUATIONS

8.1 Introduction

Any time people work together there is a potential for difficult situations to arise. In statistical consulting, these can involve conflicts between people, a problem with the project, an ethical dilemma, or a poor work environment, just to name a few. If left unresolved, a difficult situation has a way of getting worse. It certainly can impair your effectiveness.

The statistical consulting relationship seems to have a special potential for generating difficult situations. We have already explored some of the reasons why. Many of the conflicts that arise between client and consultant can be traced back to issues of customer satisfaction. Also, you have seen how the negotiations in statistical consulting can involve tangible and intangible items in fairly complex exchanges. Add to this the interdisciplinary nature of these projects and it is easy to see how misunderstandings can arise. Project-based problems are not unusual, either. Even a well-planned project can generate crises.

As you gain expertise in statistical consulting, you will probably prevent many difficult situations from arising in the first place. However, for those remaining problems, it is a good idea to know what to do about them. This chapter is about how to recognize a developing problem early, and what to do to promote a resolution.

8.2 Learning outcomes

- Characterize typical difficult situations in statistical consulting.
- Discuss how to recognize a developing problem.
- Describe how to discuss a problem with a client.
- Describe how to work towards a resolution of a difficult situation.

8.3 Breakdown!

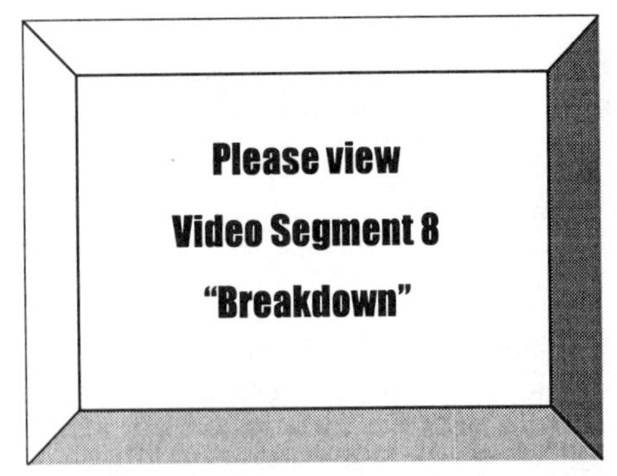

Segment 8 of the video depicts a heated discussion between Dr. Derr and Mr. Johnson. This segment takes place after Dr. Derr has sent a notebook containing univariate and cross-classified summary tables of the UMS survey to Mr. Johnson. If you can, view this segment now before continuing with this chapter. You will undoubtedly be aware of Mr. Johnson’s angry tone and Dr. Derr’s defensive responses. You may also have noticed signs that the conflict was escalating rather than moving to a resolution. The following interchange suggests the escalation:

Mr. Johnson	I knew we should have gone with those other people ... They had a much better up front understanding of what our problems were and how to solve them.
Dr. Derr	Probably a group like that was just making a lot of decisions for you that you may not have even preferred. And they probably would have charged you about double what we are charging you. Now, this has taken a lot of time ...
Mr. Johnson	Maybe you get what you pay for. I mean, this is really, I cannot use this.

This escalation is a sign that neither party is taking steps to resolve the conflict in a constructive way. Although we are not shown the entire discussion, it seems likely that it concluded with both parties feeling angry and frustrated. We can speculate that Dr. Derr might resign from the project or that Mr. Johnson might fire her.

Zahn and Boroto (1983) call situations like segment 8 a “breakdown” in the consultant-client relationship. They say a breakdown has occurred when an interaction between parties does not end to the satisfaction of all involved. These breakdowns often occur as communication problems. One reason why this dissatisfaction occurs is that either the client or the consultant had unfulfilled or unrealistic expectations. If the client is dissatisfied, he may become standoffish and hostile. If the consultant is dissatisfied, he may become frustrated and condescending. You can see this in segment 8. Mr. Johnson was highly dissatisfied with the notebook of tables. He had clearly expected a very different product. He made this expectation and his disappointment very clear in statements throughout the segment, such as:

Dr. Derr's reaction to Mr. Johnson's angry comments was to become defensive about the amount and the quality of the work that her group had done. She also attempted to explain that she needed his help in moving forward with the analysis. Unfortunately, the environment of the meeting was no longer conducive to this type of careful reflection.

Mr. Johnson	What am I supposed to do with this? ... This isn't what I asked for. ... What I had asked you for was something that I could use, something that I could take to my board, something that I could take to my committee and say this is how we can improve. Something that's readable, something that's understandable, something that's in English.

Both Dr. Derr's and Mr. Johnson's expectations of the deliverable (the notebook of tables) were unrealistic and highly divergent. Mr. Johnson expected a concise summary of how to improve the quality of services at the medical center. Dr. Derr expected Mr. Johnson to be able to peruse through 100 pages of tables and give her some insights about which ones were most important to him and how they should compute response percentages. They also were not aligned on the purpose of their meeting. Dr. Derr was unaware that her notebook had caused this reaction. She expected to have a working session about the summaries in the notebook. Instead, Mr. Johnson wanted to express his dissatisfaction. Neither party appeared to be listening to each other during the conversation. Zahn's (1988) comment about a breakdown in one of his consulting sessions seems particularly appropriate here:

> While I defended what I had done during the meeting, it was impossible for me to learn anything about what counterproductive attitude I may have had. I saw myself as an expert consultant who could not make a mistake in such an easy meeting.

While Dr. Derr was busy defending herself and her notebook, she was not able to learn about what had caused Mr. Johnson's very negative reaction. She even managed to increase his anger by showing him a diskette that contained all of the tables in electronic format! Zahn and Boroto (1983) refer to this difficult situation as a negative and downward spiral. The consultant may attempt to "kill off the client," figuratively speaking, either by using statistical jargon (the term "cross-tabulation" plays this role in segment 8), by becoming critical of the client's project, or in other ways becoming less attentive and supportive. Once this hostile environment develops, it is very difficult to change it back to a positive environment.

However, as bleak as the end of segment 8 appears, there is no reason to give up hope. Zahn and Boroto (1983) encourage statisticians to regard these breakdowns as opportunities to learn more about how to be an effective consultant. They and other statisticians have learned how to work through such breakdowns and take steps towards a resolution. One way to do this is to pay attention to signs that a breakdown might be starting. An early warning sign of an impending breakdown is the negative attributions that the consultant and client can start to make about each other. When things are not going well in an interaction, each person will often try to develop an explanation for the other person's behavior in terms of their personality, motivations, or other qualities. These explanations can often be quite negative. For example, you can imagine Mr. Johnson thinking about Dr. Derr: "She doesn't understand what I want." "She can't give me a simple answer to my question." "She has completely wasted my time and money." And, you can imagine Dr. Derr thinking about Mr. Johnson: "He can't understand these simple tables." "He doesn't appreciate all the work that went into this." "He doesn't want to take the time to help me make decisions about the data." The consultant and client may withdraw from communicating with each other because the experience has become unpleasant. Once they begin to avoid each other, they have fewer opportunities to experience each other's more positive qualities and to resolve their difficulties. They may be more inclined to argue when they do interact. Cohen and Bradford (1991) depicted this process as a negative attribution cycle, shown in Exhibit 8.1. This cycle shows that a breakdown in communication, when unresolved, can lead to negative impressions and avoidance of communication. This further reduces the chances that the difficulties will be resolved. Cohen and Bradford depict the ultimate outcome of this negative cycle as a loss of influence. In statistical consulting we can interpret this outcome to mean that the consultant has lost the opportunity to influence the client. Ultimately the consultant and client are likely to stop working with each other.

Exhibit 8.1 ***Negative Attribution Cycle***[1]

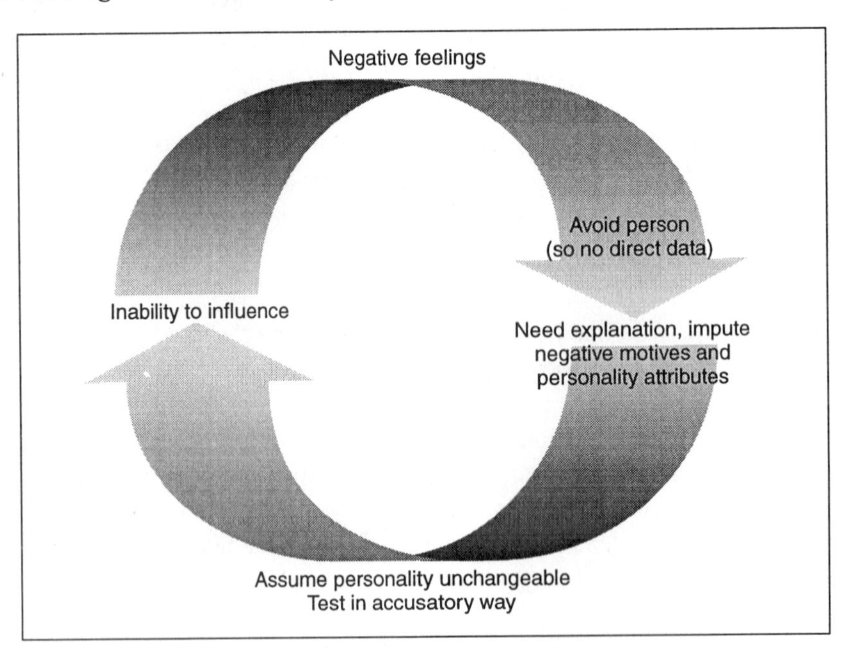

---

[1] Adapted from *Influence Without Authority*, Allen R. Cohen and David L. Bradford, p. 113. Copyright © 1991, John Wiley & Sons, Inc.. Reprinted by permission of John Wiley & Sons, Inc.

Certainly you and your client may decide that the best outcome in a given situation is to stop working together on a project. However, when that or any other decision is part of a resolution process, you and your client are more likely to be satisfied with the outcome and to leave the situation in a spirit of good will. Zahn and Boroto (1983) commented:

> All consultants with whom we have worked, regardless of level of experience, reported being concerned about their effectiveness. Furthermore, we found that there was considerable agreement between the consultant, the client, and observers of videotapes regarding the effectiveness of specific consulting sessions. What was striking to us was the causal attributions made regarding session effectiveness. Successful sessions were viewed by consultants as being a result of their expertise in statistics and their skill as consultants. Poor sessions were uniformly viewed as reflecting shortcomings of the client.

Zahn and Boroto don't report on the attributions of clients viewing these same videotaped sessions. However, we can speculate that clients might allocate blame and credit differently.

How can you avoid getting into a negative attribution cycle? Zahn and Boroto recommend that you pay attention to your internal thinking about the client during an interaction. If you discover that you are making negative attributions about this client, take this as an early sign that a breakdown might be happening or is about to happen. Then set aside these negative attributions and invite the client to work with you to identify the real difficulties. In our video story line, the problem probably originated before the conversation we viewed in segment 8. Whatever caused the divergent and mutually unrealistic expectations about the notebook was the true culprit for this breakdown.

8.4 Resolution

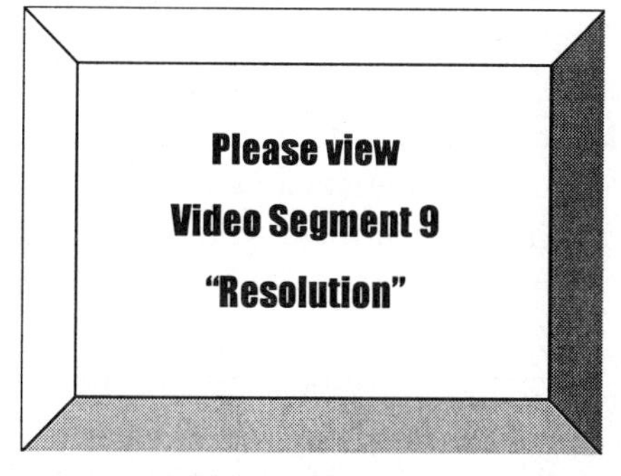

The next segment of the video story line, segment 9, finds Dr. Derr and Mr. Johnson at a coffee shop. If you are able to view this segment now, do so before continuing with this section. Dr. Derr selected a neutral location with a more relaxed atmosphere. She invited Mr. Johnson to have a cup of coffee with her and talk about the project and their previous meeting. Dr. Derr began with an apology about her unrealistic expectations regarding the notebook. She explained that because she enjoys looking at data and analyses, she had not really considered the impact of such a notebook from Mr. Johnson's point of view. She identified what had happened in segment 8 as a "consulting breakdown" and explained that breakdowns often happened when people in two different disciplines worked together. Mr. Johnson also apologized for his angry comments. He revealed that he had developed his expectations about the notebook from a report that had been sent to him. This report was the final product of a similar survey that had been conducted at another institution. This discussion helped to reveal how each person's expectations had set them up for the frustration, anger and defensiveness we saw in segment 8.

In segment 9, Dr. Derr provided a summary table and two charts to show Mr. Johnson what she wanted to develop from the survey analysis. This material is shown in Chapter 7, as Exhibits 7.10 and 7.11. She wisely selected summaries that addressed the key objectives of the survey. Mr. Johnson's enthusiasm for the table and charts was evident in the segment. He immediately began to interpret the results in terms of the survey objectives. Dr. Derr was then able to point out how the table of demographics (Exhibit 7.10) and the pie chart had been developed from one of the summary tables in the 100-page notebook (Exhibit 7.9). She could show Mr. Johnson what additional input she needed from him in order to finalize the chart. Seeing the end product and its relationship to the data summaries helped Mr. Johnson understand better what Dr. Derr needed from him. Dr. Derr appeared to be equally glad to have a model of a report from a similar survey that showed what Mr. Johnson wanted to have as an end product. Segment 9 represents the meeting that Dr. Derr and Mr. Johnson probably should have had, rather than the meeting in segment 8. Both parties were clearer about their expectations. Mr. Johnson provided a model report. Dr. Derr produced some preliminary charts. With the end product more clearly in view, they could then work together to make the decisions that would lead to that product.

8.5 Conflict Resolution

The video story line in segments 8 and 9 suggests that it would be a good idea for a statistical consultant to learn some skills in conflict resolution. These skills will help you to break out of a negative attribution cycle and get back on track with a satisfactory consulting relationship. In segment 9, Dr. Derr illustrated some of the skills required in conflict resolution: (1) She openly discussed the problem; (2) She framed the problem in neutral terms so that neither consultant nor client felt "blamed"; (3) She invited Mr. Johnson to describe what had been making him dissatisfied; (4) She listened to him; (5) She was willing to acknowledge her contribution to the breakdown. This approach helped to set the stage for Mr. Johnson to reciprocate. He refrained from the accusatory tone that he had adopted in segment 9. He also acknowledged his contribution to the breakdown. He was then willing to make an effort to get the project back on track. He listened to Dr. Derr while she described the source of her dissatisfaction. Then together they figured out how to move forward in such a way that both of them would be satisfied with the outcome.

The resolution of a breakdown does require cooperation between the consultant and the client. Although we can't order a client to cooperate, we can do a lot to enhance the chances that this cooperation will take place. First, we need to make sure that the outcome of the project will make both (or all) parties satisfied. This is the basis of the "win-win" negotiation that we discussed in Chapter 6. Then, we need to stay in communication with the client. This will help to build up trust so that an occasional "blow-up" will not derail the project. Finally, we need to make sure that the way that we react to problems is one that will elicit a cooperative, problem-solving response from the client and other members of the project team. By identifying an emerging breakdown early, discounting negative attributions, and then discussing the problem in neutral terms we can help to minimize the harmful consequences of these breakdowns.

Exercise 8.1

Put yourself in the place of the statistician in the video story line. Describe what you might have done during the meeting shown in segment 8 to do the following:

a) Recognize that a breakdown was taking place;

b) Minimize the impact of the breakdown during the meeting;

c) Promote a resolution of the breakdown during the meeting.

Exercise 8.2

Below is a description of a consulting project in which a breakdown took place.

Trouble in the Corner Office

Jack Wood† and Samantha Davidson were faculty members in the business school of a large state university. Both had interests in business forecasting, and they were delighted to receive a consulting contract with Stocks-R-Us (SRU), a growing company that specializes in financial services for small and medium-scale investors. The project involved the development of a comprehensive forecasting system for the stock market, with an emphasis on the securities that SRU traded most heavily.

Sam and Jack worked with Simon Johnson, a bright young executive of SRU and reported to Simon's boss, Graham Bridges. Graham has an MS in Statistics, which he received about twelve years ago, and of which he is very proud.

Although he has mostly been an administrator for the last seven years, Graham feels he has kept up his statistical knowledge rather well. The project went rather well and Sam and Jack were able to produce the first model according to the agreed schedule. Simon suggested that they present the details to Graham, which they duly did at a meeting of the four of them. Although Sam and Jack felt that the presentation went rather well, Graham seemed to have some difficulty in following the rationale for the model, as well as some of the more technical details. At the end of the meeting, Graham proposed that he recruit

---

† The names and details of examples appearing in this chapter, unless otherwise indicated, are fictitious.

two outside experts to appraise Sam and Jack's model, a suggestion they readily accepted.

Two weeks later, in a seminar-like setting, the original foursome met with the two experts and spent a fascinating hour and a half discussing the issues. Sam and Jack enjoyed the interaction with the two experts, who declared themselves well satisfied with the overall model, while making several valuable suggestions for improvements in certain areas. Simon remained respectfully quiet during the meeting; Graham seemed uncomfortable and out of his element.

The project wrapped up successfully a few weeks later. However, relations between Sam and Jack, and Graham, deteriorated. Although Simon was able to implement the model successfully and SRU continued to prosper, Graham terminated the consulting relationship with Sam and Jack and they were unable to publish details of their work in an academic journal.

1. Describe the breakdown that occurred during the course of this project.
2. Why do you think the breakdown took place?
3. Why do you think the breakdown had the effect that it did?
4. What could Sam and Jack have done to address the breakdown at the following points in the course of the project:
    a. To anticipate and attempt to avoid the breakdown in the first place;
    b. To recognize and address the breakdown while it was taking place;
    c. To resolve the breakdown after it had occurred.

8.6 What Type of Difficult Situation Is It?

The video story line presented one type of difficult situation, and in some respects it was a type that could fairly easily be resolved. The scope of the problem was "short-term" because it did not extend beyond the project itself, and was actually limited to one phase of the project. Dr. Derr and Mr. Johnson could resolve the problem between themselves without involving anybody else. Nothing had gone wrong with the statistical parts of the project. The source of the problem was a lack of communication about what Dr. Derr and Mr. Johnson should expect from each other at this stage of the project. They were able to resolve the problem because their positions were readily negotiable. Even though Mr. Johnson got angry during segment 8, lack of professional behavior did not really contribute to this problem. Exhibit 8.2 is a summary of the attributes of the difficult situation in the video story line. Included with this summary are contrasting attributes of other difficult situations that a statistician might encounter. For example, a difficult situation might involve a whole project team instead of just one client. The scope of the problem might be "long-term" if it extends beyond the actual project and encompasses issues of policy or involves other people. The problem might originate with the project itself, such as problems with design, data collection, analysis or interpretation. The positions held by the people involved might not be negotiable. Finally, some difficult situations are caused or at least made worse by unprofessional behavior on the part of one or more of the parties involved.

As a statistician, you may have already calculated that if five attributes are needed to characterize a difficult situation, and each attribute has two contrasting levels, then there are $2^5 = 32$ types of difficult situation that you might encounter and need to resolve in your consulting practice! This variety is certainly one reason why statistical consulting contains enough challenges for a lifetime of learning. In fact there are probably more than 32 types of difficult situations as other factors get involved and interact in complex ways. However, it is useful to look at the five attributes in Exhibit 8.2 and consider how you would respond to situations that are quite different from the one described in the video story line.

Exhibit 8.2 ***Types of Difficult Situation***

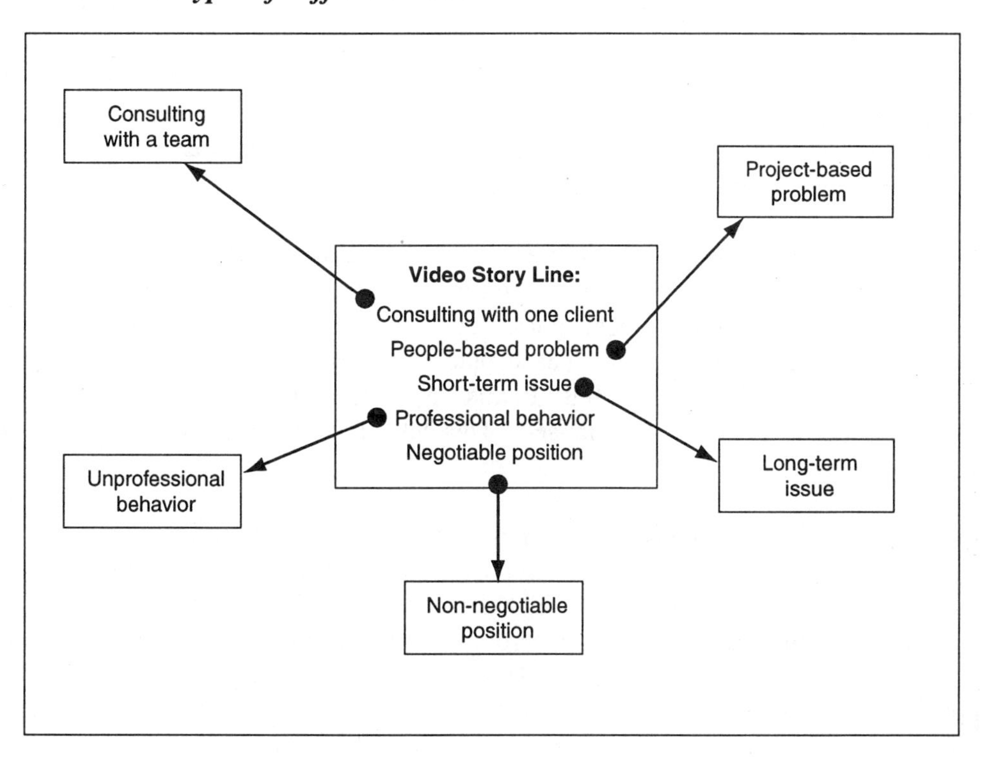

8.7 A Long-Term Issue

Let us return to the consulting story in Example 6.1. Nathan Thomas was the statistician who worked with a "go-between" for a physician client, Dr. Picardo. Nathan did a lot of work for Dr. Picardo's project without really knowing if he was doing what he was supposed to do, or even how much he should charge for it. This situation was fairly representative of much of the consulting that Nathan did with physicians where he worked. In fact, this part of his job had become so frustrating and unrewarding for Nathan that he was seriously thinking about quitting. However, he decided to make one more effort to see if he could improve things at work before polishing up his resume and sending it out. Let us take a further look at this story to see what Nathan did to improve the outcome of his interactions with the physicians where he worked:

The causes of Nathan's difficult situation extended beyond the specific interaction between him and Dr. Picardo on one project. It was necessary for Nathan to address policies and attitudes that were at the root of the problem. This is what makes this a "long-term" problem. The first thing Nathan did was to identify the consulting issues that contributed to the problem. As you read in Chapter 6, these are: Issue #1, Nathan's role; Issue #2, the role of others on the project (especially the go-between); Issue #3, communication; Issue #4, deliverables; and Issue #6, compensation. With each issue, he identified what he would consider to be the "ideal" consulting situation. He decided that he wanted to be able to work more directly with the physician rather than with a go-between. He wanted to have a clear idea of what the physician expected him to provide for a project. He also wanted to have a clear understanding with the physician about how his efforts on the project would be charged, before doing any work. On further reflection, he decided that it was no longer acceptable to him to work entirely with a "go-between" with no access to the physician in charge of the project. If he were not able to change this arrangement, then he would in fact look for another job.

What Nathan has done at this point was to step back from the immediate problem and think about what factors were contributing to the problem. He then identified his position. He described what to him would be an ideal situation, and he also described what would be unacceptable. Between the ideal and the unacceptable situation was a range of possibilities that he could accept depending on negotiations. In this way he gained perspective on the problem and set some goals and boundaries for himself. This helped him to evaluate different possible strategies.

Because there are so many factors that come into play in a work situation like this, there will be no one correct thing for Nathan to do that will produce one guaranteed successful outcome. To illustrate this variety, we are going to "clone" Nathan and place duplicates of him in two different versions of the story. In exercises throughout this chapter, you are invited to create your own story lines for difficult situations. This will give you further practice in considering strategies and outcomes as you approach the difficult situations in your work environment.

> **Nathan Clone #1:** Nathan telephoned Dr. Picardo and described his frustrations in providing statistical support for his projects. He asked to be able to meet with Dr. Picardo directly to discuss the statistical needs of a project and the charges involved. Dr. Picardo admitted that he had not been satisfied with some of the statistical work and had not understood the need for some of the charges. However, he said he did not have time to attend a lot of meetings. Nathan replied that actually Dr. Picardo would end up saving a lot of time by making sure that Nathan understood the project and what the highest priority statistical needs were. He suggested that he, Dr. Picardo, and Josef Martin (the go-

between) have one organizational meeting before any analysis was done. After the meeting, Nathan would write up an estimate of his charges and send it to Dr. Picardo. Then Nathan and Josef could work together, with the option of calling in Dr. Picardo for any critical decisions. Dr. Picardo agreed to this. Nathan hoped that, as Dr. Picardo began to experience greater satisfaction with this arrangement, he might be open to other suggestions that would improve Nathan's input on his projects.

Note that Nathan took his cue from Dr. Picardo's comment about time that this was an important factor to him. He then framed his argument for a planning meeting in terms of the overall savings of time. Nathan was also willing to compromise with Dr. Picardo in terms of the number of meetings they had together. In this version of the story, the improved communications set the tone for future improvements.

Nathan Clone #2: Nathan was unable to reach Dr. Picardo by telephone. Instead, he set up a meeting with Josef Martin, the go-between. He described his frustrations to Josef and learned that Josef was equally frustrated. According to Josef, Dr. Picardo was very busy and traveled a great deal. Much of the day-to-day work on the research studies was left to Josef. However, Josef's level of authority in these projects was unclear. Dr. Picardo still wanted to make decisions about the research, even though he was often unavailable during critical periods. Dr. Picardo did not appear to understand the statistical work that Nathan produced, and he would become angry with Josef if the results from the studies were not what he had expected them to be. Josef did not really understand the statistical issues either, and he tended to pass the blame on to Nathan. Josef seriously doubted that Dr. Picardo would meet with Nathan to discuss the research. Apparently Dr. Picardo perceived both Josef and Nathan as possessing a lower status than himself. He generally met with and took suggestions only from individuals whom he perceived to be of equal or greater status.

After this conversation, Nathan offered to work with Josef and explain the statistical issues of the research studies more clearly to him. He suggested that they meet at the planning stages of a study and then regularly after that, so that Nathan could provide input all along. He also suggested that they create detailed project documents, and that both of them document their activities, decisions, and attempts to reach Dr. Picardo. Whenever Nathan provided any statistical materials, he made certain that Josef understood them.

Meanwhile, Nathan began to work in an improved way with the other physicians who were more open to meeting with him. He ordered his workload so that these projects took priority over Dr. Picardo's work. He arranged with the head of the unit to have a well-known statistician visit the group and give a series of talks. He talked with the statistician beforehand to make sure that she included plenty of horror stories about the consequences of not working with a statistician at the planning stages of research. He also volunteered to serve as a statistical reviewer for the premier medical journal in Dr. Picardo's field.

In this version of Nathan's story, lack of clarity about the roles of people on the project and the importance that Dr. Picardo placed on status both contributed to the difficulty of the situation. For the immediate situation, Nathan tried to contain the damage from the blame that was assigned to him. He also made sure that he and Josef worked well together. The amount of documentation that he suggested was another protective measure. Meanwhile, he took his cue from the fact that Dr. Picardo clearly valued status. While it is true that no amount of negotiation and clear communication can change a person's perceived status, Nathan still was able to make use of status to influence Dr. Picardo. He cultivated his relationship with more approachable colleagues of Dr. Picardo. This would help protect him against any blame that Dr. Picardo might send in his direction. These colleagues were also in a better position to let Dr. Picardo know that their research had been improved by meeting directly with Nathan. Someone who values status will often adopt practices that he sees his peer group adopting. Nathan had Dr. Picardo's boss invite a high-status statistician to give talks about the value of involving a statistician in the planning stages of research. He also increased his influence by serving as a statistical reviewer for a journal to which Dr. Picardo ascribed a high status. All of these are indirect, long-term strategies. They may or may not balance the risk of damage that Dr. Picardo's poor research practices represent to Nathan. Depending on how the situation plays out, Nathan may or may not be polishing his resume. However, the long-term actions that Nathan took are all positive measures. His overall working environment may improve so much that Nathan may decide he can live with Dr. Picardo and his research projects.

Exercise 8.3

Return to the story of Nathan, Clone #1. Suppose that Dr. Picardo told Nathan that his research funds were limited, and he could not afford to pay for Nathan's time for a lot of meetings. How should Nathan frame his arguments in this case? Describe some options that Nathan could suggest to Dr. Picardo.

Exercise 8.4

N. Philip Ross (1995) related the following description of a difficult situation in a government agency:

In particular, upper management in the government have not the foggiest idea what is going on below them in terms of the information needs that they have. Statisticians need to understand that. As an example of this at EPA [the Environmental Protection Agency], a few years back one of the assistant administrators for water gave a talk at a statistics conference. She was known for not having much use for statisticians, so we invited her to find out why. It was not a milieu of animosity or antagonism. She liked us, but just did not see what we did for her. One of her reasons she gave as an example. She had to advise the Army Corps of Engineers on whether or not they should be allowed to build a dam on a river relative to some environmental impact concerns. ... This advice had to be given to the Army Corps of Engineers in the next three months, and they would then act on it. Since statisticians obviously deal with data and provide information, she brought in statisticians from her staff and asked them to provide her with as much information as they could so she could make that decision. The answer from the statisticians was that the data that had been collected were not very good, the monitoring systems were set up to measure for compliance to the law, not with regard to the state of the environment or the impact that might occur. They said if they were given a couple of million dollars, and about two years, they would set up a survey design and monitoring system, and would come back and tell her what to tell the Army Corps of Engineers. This was not of much help to her, of course, because she had to make the decision in the next three months.[2]

1. The statisticians and policy makers already seem to have developed some negative attributions about each other. Describe what these negative attributions are. What do you think is the root cause (or causes) of the misunderstandings that has led to these attributions?
2. Suggest some changes in the situation that the statisticians could initiate that would improve their effectiveness. Identify some small changes that could be implemented fairly readily and others that are larger in scope.

Exercise 8.5

The following are descriptions of typical difficult situations for statistical consultants. For each one, suggest what the statistician could do to improve the situation. As in Exercise 8.4, identify some small changes that could be implemented fairly readily and others that are larger in scope.

1. My client tells me that "everyone in my field does it this way". I want to suggest a statistical method that I prefer.
2. At present, the statisticians at my work place are not included at all stages of work. I am often asked to provide an analysis of data that is just dropped on my desk, without any opportunity for involvement.
3. There is a great deal of skepticism where I work about statisticians. People don't seem to trust us or give us much credibility.

---

[2] From "What the government needs" by N. Philip Ross, p. 8. Copyright © 1995, American Statsitical Association. Reprinted with permission.

8.8 Consulting with a Team

Another factor that can contribute to the difficulty of a situation is the number of people involved. An interdisciplinary team can address the diverse needs of a project very effectively. However, disharmony and conflict among the members of a team can degrade productivity. When conflict arises, team members may spend a lot of time and energy on disagreements and focus less on the actual work. Members may also begin to avoid each other as the interactions become more and more unpleasant. They may blame other team members. The team may fragment as members "choose sides". Members may start to become less interested in participating in the project. This will make it difficult for team members to make decisions and implement actions.

Conflict itself is not a bad thing for a team. When handled respectfully, disagreements among team members can reveal different perspectives that all contribute to the creativity and problem-solving capacity of the team. Yeatts and Hyten (1998) call this "cooperative conflict." When two or more team members have opposing opinions about an issue, they can approach this difference in a spirit of exploration and understanding. When they discuss their differences, they can expect that the resolution will produce a more robust solution to the team's problem. This is a "win-win" approach to conflict. Cooperative conflict is essential to a team. Without it, team members can begin to think alike and to accept each other's ideas uncritically just for the sake of harmony. This team phenomenon is sometimes called "groupthink." As you can imagine, a team that does not expose its members' ideas to critical review is not likely to address the problems of a project very realistically.

In contrast to cooperative conflict, Yeatts and Hyten identify "competitive conflict." Members engaged in competitive conflict will vigorously defend their positions and try to win others over to their point of view. They will do everything they can to undercut their opponent's position. They may try to use their authority or their status to further advance their position. This "win-lose" approach to conflict will reduce the cohesion and trust among team members. The project is the biggest loser because team members are no longer motivated to work together and give their best efforts to meet its goals. How can you tell, in the midst of a discussion, whether a team is engaged in a cooperative or a competitive conflict? People can argue heatedly with each other while still maintaining respect. People can also appear calm and civilized while they engage in nasty behind-the-scenes tactics to undercut each other. The best way to tell the difference between a cooperative and a competitive conflict is to examine the outcome. How do the team members feel about the outcome? Do they have hard feelings about the other team members? Is there someone who feels that they "lost" the argument? If so then this was probably a competitive conflict. Do they accept the resolution and look forward to the next discussion? If so then you can be pretty confident that this was a cooperative conflict.

What can a statistician do to promote cooperative conflict and reduce competitive conflict in a team situation? This will depend a lot on the nature of the team and the role of the statistician on the team. To get some ideas, let us look at several examples:

Example 8.1

Frank is a statistician who works for a company that manufactures athletic clothing. He met with a team that was responsible for testing a new fabric. The team included market analysts, textile engineers and a physiologist. The team had been assembled to evaluate the comfort and heat resistance of several versions of the new fabric. Frank's tasks were to design the assessment studies and then to analyze the data.

Frank's first impression of this team was that the members did not like or trust each other. Discussions were awkward. Nobody seemed willing even to make eye contact with anyone else. After the first meeting, one of the engineers took Frank aside and complained that the marketing people were pressuring her to endorse the performance of the fabric without adequate testing. Later, one of the market analysts complained to Frank that the engineers were constantly tinkering with the fabric, and that the physiologist's proposed tests were too expensive and time-consuming. The physiologist also complained to Frank that he was unable to complete a test run of the fabric because of all of the changes that the marketing and engineering members made. The team leader, the senior market analyst, expressed her doubts to Frank that the team really needed input from a statistician.

What should Frank do? This does not sound like a cohesive group. The members were engaged in competitive conflict outside of the team meeting. They all attempted to get Frank on their "side" by blaming the other team members for the team's lack of productivity. Their positions may be entrenched and difficult to modify. The team leader did not even make Frank feel that his contribution would be valued. In this situation, Frank may be at risk professionally. It would be easy for everybody to begin to blame him for the team's failings. It would be difficult for Frank, acting on his own, to improve the functioning of the team.

Frank can get some perspective on the situation by finding out more about the decisions that led to the formation of this team. In doing so, he needs to be careful to learn about the background of the team without engaging in office gossip. He may be able to find some assistance elsewhere in the organization to intervene and improve this team. Frank can also try to work with the team to improve its cohesiveness. He is in a good position to listen to each team member's point of view about the technical problems of the project. However, he should avoid taking sides with any one person or subgroup. He may then be able to facilitate an increased understanding among team members of their different perspectives. He would also be wise to take steps to protect his own professional interests within the organization from the consequences of being associated with such a dysfunctional team. He may also want to consider the option of leaving the organization if little support is given to him and his interaction with these teams.

Example 8.2

Micki is a statistician for a survey research organization. She was assigned to work with a team of social scientists to develop a survey. Her tasks were to develop a sampling scheme, help the team develop the questionnaire, oversee the data management, and then analyze and interpret the results. She had the specialized knowledge needed to accomplish these tasks. However, the other members of the team perceived Micki as abrupt and arrogant. At team meetings she used a lot of technical jargon to discuss the sampling plan and the analysis. When somebody asked her to clarify some of the concepts, Micki had a way of making that person feel stupid for asking.

Eventually the rest of the team began to withdraw from Micki. They made decisions about the questionnaire without her. They had fewer formal meetings and instead carried out much of their business in the lunchroom, while Micki dined alone in her office. They began to work informally with Jason, a sociologist who had joined them in the lunchroom. Jason had some statistical training and expressed interest in the analysis of the survey. On Jason's recommendation, the team requested specific parts of the database from Micki. These requests were in the form of email notes to which Micki responded by emailing the data files. With Jason's help they analyzed the data on their own. Micki, meanwhile, complained to her friends that the survey group at work refused to listen to her advice. She said that she stopped being invited to team meetings. She complained further that the team only asked her to do computer programming and data handling, and did not make use of her expertise in data analysis. She was particularly angry when she learned that the group had consulted Jason about the data analysis. Micki knew that Jason had much less expertise than she did. She was sure he wasn't incorporating the sampling weights correctly. She concluded that social scientists were impossible to work with.

This team essentially "fired" Micki and "hired" Jason. Can you understand why? Although Micki had the necessary expertise for the statistical tasks, her insensitive manner made it unpleasant and difficult for the rest of the team to make use of it. Micki's negative attributions about the social scientists may help her bruised ego but they do not help her to become a more effective statistical consultant. Instead, she has withdrawn from the team and they have separated from her. If Micki spends too much time behind her office door, chances are that her contribution may be viewed as increasingly marginal to the organization. She may lose her position.

Example 8.3

Sean is a statistician for a pharmaceutical firm. He started working with a research team on a product that affects milk production in dairy cows. The team included animal scientists, veterinarians, and chemists. The team leader was from the regulatory affairs group. Until recently in this firm, members of the statistics group had not been included in product development teams. Instead, the statisticians were usually consulted at the last stage of protocol development to get a sample size and again after the data had been collected to get a p-value. Sean wanted to participate more actively with this team and to be seen as a valuable team member. In addition to having a graduate degree in statistics, he also had some training in animal science. For this reason, he felt that he had a lot to offer to this team. From informal conversations at work, he realized that statisticians in this firm had a reputation for demanding unrealistic sample sizes and for producing complex data analyses that nobody could understand.

Sean started his campaign by getting to know each team member individually. In various informal settings (mostly in the hallway and in the firm's cafeteria) he chatted with different members and found out more about their views of the project and their views (not always complimentary!) of statistics and statisticians. He tried to find things that he and the other members had in common. He also asked for resources from different members that would help him understand their perspective of the project. This is where his training in animal science was useful. Through these informal conversations, the other team members began to realize that Sean could appreciate the scientific issues of the project. They began to trust that he would keep the issues and constraints of the study in mind when providing his statistical input.

Sean made sure that he heard about and attended all of the project meetings. At the first one, the team leader put the statistical issues of the protocol first on the agenda, and then told Sean that he could leave after that. Instead, Sean asked to stay, since he felt there was a lot that he could learn. The team leader declared that this was the first time that a statistician had ever asked to stay at a meeting!

When it came time to discuss the study design and sample size, Sean prepared a graph and a table that illustrated the statistical power of the range of sample sizes under consideration. He showed drafts of these visuals to several team members before the meeting to get their feedback. This helped him to present his recommendations clearly to the team. He worked in a similar way at the analysis stage, showing preliminary drafts of results in various formats to several members to see which way of depicting the results was the most informative.

Sean's work with this project suggested several statistical topics that would recur in future projects. He felt that the team would benefit from knowing more about these topics. He offered to present series of talks on these topics, making use of the data from the team's previous work.

Why was this a difficult situation? The previous reputation of the statistics unit, and the past history of statisticians who did not attend meetings set up some negative expectations from Sean's team members. Sean counteracted these expectations and avoided problems before they could arise. He took many steps to reach his goal of being viewed as a valuable team member. It would be difficult for the rest of the team to miss the message that he was interested in the project and committed to the team's goals. He also made sure to establish a good rapport with individual team members. Once the other members accepted him on the team, he was able to develop some long-term strategies that would help this team improve the statistical practices in future projects. Now Sean faces a new challenge, persuading the other members of his statistics unit to adopt these effective practices.

Example 8.4

Lisa Dumfries is a statistician at an academic consulting unit. She was head of an assessment team for an educational research project. Her team included Dr. Vijay Singh from the statistics department, Dr. Chia Lee from the educational psychology department, and Olga Metzger, the owner of Edu-Tech Inc. The task of this team was to develop and test an assessment instrument that could be used to evaluate different types of computer-based training. The project was funded by a government grant (awarded to Lisa) for university-business initiatives. Lisa was aware of the tensions that existed between the two academic departments and between the university and the small businesses in the town. She set out to make sure that these tensions would not derail the project. Prior to their first team meeting, she met separately with Dr. Singh, Dr. Lee and Ms Metzger. She wanted to learn more about what each person wanted to accomplish from this project, and what their concerns were. Both Dr. Singh and Dr. Lee wanted to publish from the results of the study. They each had students assigned to the project. Olga wanted this project to help establish her firm's position in the computer-based training market. She was concerned about meeting the project deadlines and staying within the budget. While sympathetic to the need for publications, she wanted to make sure that proprietary information about her company's computer programs would remain confidential.

Lisa felt that establishing good communications would help this team avoid some of the pitfalls that could threaten the project. Finding out that all of the members were comfortable with email, she set up an email list-serve to facilitate communication among the team members. She selected a hotel conference room in town as a neutral location for their first meeting. Subsequent meetings were rotated around the four locations of the principal team members. She established a schedule for telephone conference calls to keep the team in touch between meetings. Lisa also addressed the substantive concerns of the team. Early on in the project, she initiated a discussion about publication and confidentiality issues. She asked Olga to identify the proprietary information from her firm. She then worked with the university administration to develop an agreement that would protect this information while providing rights to publish.

At the first meeting, an argument broke out between Dr. Singh and Dr. Lee about a point of methodology. The argument reflected the long-standing tensions between their two departments. Soon their students chose sides and joined in. As the argument became more heated, Ms Metzger and her staff were left on the sidelines to wonder at the polarization between the two departments. Lisa intervened. She reminded the group of the purpose of the meeting and the goals of the project. She suggested that the team set up a regular bimonthly meeting to review literature and discuss their perspectives of the methodology. She deferred the decision about the point of methodology until a later time.

This bimonthly seminar was the first cooperative venture between the two departments. The discussion was very lively and attracted many participants. Lisa moderated the discussion and made sure that it was conducted respectfully.

The team project was finished successfully and had several other beneficial outcomes: Several students proposed research projects that spanned the two departments. Edu-Tech developed an internship program with the university. And, as the principals became better acquainted, they were able to suggest ways that Edu Tech could improve its market position.

Why was this a difficult situation? The project brought people together in a cooperative venture who had a past history of conflict and tension. The statistician in this example was in the role of team leader. As such, she had a lot of influence on the collaborative spirit of the team. She used this influence to defuse the tensions between the groups. You can think of these tensions as sets of negative attributions that each group had about the other groups. These attributions had probably accumulated over time and at least in part represented the different interests and perspectives of each group. In order to counteract these negative attributions, Lisa had to set up good communications, encourage team members to get to know each other, and to address the needs and concerns of each participant. When a breakdown occurred, Lisa had to intervene to defuse the argument. She did not take sides or deny the existence of differences. Instead, she created a more appropriate means for the differences to be discussed. The positive outcomes of this project illustrate the benefits that can come from a collaborative interaction.

Exercise 8.6

Suggest some ways that Frank (from Example 8.1) could facilitate a greater appreciation of each team member's perspective on the project.

Exercise 8.7

Examples 8.2 and 8.3 both mentioned casual interactions in the office lunchroom and hallways. What part does this type of conversation play in preventing difficult situations?

Exercise 8.8

Suppose that you are Micki's supervisor (from Example 8.2). You have learned about her separation from the survey team and you have heard about some of her negative attributions. What would you say to Micki? What could you do or provide to help her work better with these teams?

Exercise 8.9

Suppose that Sean's (from Example 8.3) work with the dairy cow team produced glowing reports from his team members. On hearing this feedback, the rest of the statistics group found a lot of reasons why this experience was unusual and would not generally apply to the work that they did. They remarked on how much extra time Sean spent with this team. They continued to complain a lot about the non-statisticians they work with. How could Sean help his colleagues to have more positive experiences with their clients? What could the leader of the statistics group do?

Exercise 8.10

Return to Example 8.4 and focus on Olga Metzger's concern about budget and deadlines. Suppose that she has the impression that academic faculty often use up a lot of funds without providing a useful product, and that they are likely to give higher priority to their work in theory than to their work in applications. If you were the team leader, how would you address her concerns?

8.9 A Project-Based Problem

Sometimes a difficult situation does not arise from the interactions among the people involved. Instead, the project is the problem!

Anything can happen during the course of a project. You may discover that the design of the study was flawed and it is too late to change it. Some accident or other event may result in the

loss of data. Perhaps you or one of the other team members made an error. These are all difficult situations because they can adversely affect the goals of the project.

Suppose that you are the one who has identified a problem. How should you discuss it with your client? Here are some guidelines:

1. Identify who should hear the "bad news" first. This should usually be the person or persons who are ultimately responsible for the project. If they hear about the problem indirectly, or after everyone else on the project knows about it, or from the morning newspaper, they may justifiably become angry and embarrassed. This will not help reach a resolution of the problem.

2. Choose an appropriate setting for the discussion. If possible, this discussion should be in person. It might seem much easier just to send the person in charge a memo rather than to face them in person. However, a discussion of a problem can be sensitive and complicated. A face-to-face discussion is better suited to deal with complicated communication. If this is not possible, then try to have a conversation on the telephone. A written report on the problem and its follow-up will then be a useful way to augment your discussion.

3. Use appropriate language in discussing the problem. Nobody likes to feel that they are being accused of something. You won't make much progress in resolving a problem if everyone acts defensively. Instead, focus on the problem rather than on the person. You can do this through your language. Exhibit 8.3 shows some examples. The positive version of all three statements in Exhibit 8.3 makes use of "I" language. With "I" language you can make it clear that you are expressing your opinion about the problem. This sets a neutral, non-accusatory tone to the conversation. In contrast, the negative version of all three statements is accusatory. The "you" in the statements places the blame squarely on the client. On hearing the negative version of these statements, the client is likely to feel defensive. He may try to find a way to blame you or someone else instead. This does not set the tone for a problem-solving discussion.

In addition to the "I" language, there are other constructive elements to the positive version of each statement. Each positive statement provides a lot more information about the problem than does the negative version. Statement #1 in Exhibit 8.3 suggests how the problem might ultimately be resolved (by including limitations to the conclusions drawn from the study). Statement #2 also suggests a solution (unscrambling the information about the labels) and engages the client in the process of finding the solution. Statement #3 expresses respect for the client's approach and suggests options for deciding on the best model (comparing the two models, bringing in an outside mediator). This statement also anticipates for the client that the conclusions from the more appropriate analysis will be different from the ones that he obtained.

4. Express your concerns in terms of the goals of the project. If you simply declare that a study design is flawed, or that an analysis is invalid, you have still not connected your statistical concerns with the objectives of the study. Your client may dismiss your comments if he feels that they relate more to some arcane issue in statistics rather than to his interests. You will have

Exhibit 8.3 ***Positive and Negative Versions of "Delivering Bad News"***

	Instead of saying this...	Try saying this ...
1.	Your study was poorly designed.	I am concerned about the way the study was designed. I believe there will be limitations to the conclusions you can make about ...
2.	Your technician messed up all the labels in the latest blood collection.	There was a problem in the latest blood collection that we will have to unscramble. I will need your help ...
3.	Your analysis model was incorrect and the conclusions you have made are invalid.	I learned a different way to analyze the data. I'll be glad to discuss the analysis with you and show you what the differences are between your and my approach. If you like, we can bring in someone with more experience to help us decide which model is more appropriate in these circumstances. Since this will probably affect the conclusions we draw from the study, we need to make sure to use the most appropriate analysis.

more impact if you translate your concerns into terms that relate directly to the client's interests and the objectives of the study.

For example, let us return to Dialog 6 in Chapter 5. In this dialog, the consultant has learned that the client intends to give 12 pigs a low fat diet in the first month of the study, followed by a medium fat diet the next month, followed by a high fat diet in the third month. Now, let us suppose that you have learned about the study after it was finished. There is nothing you can do about changing the design at this point. Instead, you would like to express your concern about the inference that can be drawn about the diets. You could simply say to the client "This design was flawed. You can't determine a valid effect of diet because it is confounded with month." However, your client may not see the connection between month and diet in the same way that you do. You can initiate a more constructive approach to the problem when you explain and explore this connection. Exhibit 8.4 contains the statistician's part of a more constructive dialog.

The statistician in Exhibit 8.4 spelled out the consequences of the confounding in terms that were important to the client: if the study's findings could not be replicated by others, this could reflect poorly on the client and on the research unit. She then invited the client to consider options that would help protect the client from these consequences.

Exhibit 8.4 ***Expressing Your Concerns in Terms of the Goals of the Study***

> My understanding from our discussion to this point is that the same order of test diets was used for each pig.
>
> . . .
>
> I can appreciate all the labor that went into this study. I guess it made it easier to be able to make up just one diet formulation each month.
>
> . . .
>
> I have a concern about this approach when it comes to interpreting the results. If the diets were always given in the same order, then we can't make the distinction between the effect of the diet and the effect of month. This is sometimes called "confounding." There might be some differences between the three different months of the study that we can't distinguish from the three different diets. This can be a problem. Someone might read about the results from this study and try to apply the diet that was recommended. If the apparent effect of the diet actually came from the effect of the month, then others may not get the same outcome. If a lot of people have the same experience, that might come back to this research center in an unfavorable way.
>
> . . .
>
> What do you think might have an impact on the responses that you measured? I know that sometimes season has an effect on metabolism. And sometimes there are trends in the way the study is carried out over time. Why don't we brainstorm about what the impact of some of these factors might have?
>
> . . .
>
> Well, based on our discussion, I think there are several options that we can consider. I think it is important that whoever reads about this study knows about the order and the months in which the test diets were given. This can help them make their own judgment about how applicable the results will be to their situation. You could summarize the effects of season that we have just discussed, with some references. We could include some of the environmental measurements you made to support your belief that the most important environmental factors did not change during the period of the study. And I'd like to help you with the wording of the statistical conclusions. I want to make sure that readers understand that the effect of each diet is conditional on being measured in a certain month and in a specific order. What do you think about these suggestions?

5. Balance the negative with something positive. It can be discouraging to hear bad news. Try to find positive things that you can also say about the situation or about the project in general. This approach helps many people deal with bad news more constructively. In Exhibit 8.4, the consultant acknowledged the amount of labor that went into the study. Making positive statements helps create empathy. This will show the client that you are willing to help resolve the problem.

6. Offer options. Providing options is part of the problem-solving process. When you deliver bad news with good grace, you and your client can get beyond an emotional response and back to the business of figuring out what to do about the problem. In Exhibit 8.4, the consultant offered several options, and then invited the client to give his opinion about them.

7. Take responsibility for your errors. Oops! What if you are the one who made an error? Well, even very experienced statisticians make mistakes once in a while. The important part of making an error is taking responsibility for it. What does this mean? If you follow the example of some political figures of our time, you might be tempted to cover it up, lie about it, and then when cornered with the evidence, deny the error with convoluted reasoning such as "It really depends on what your definition of 'significantly less than' is." Actually, taking responsibility means admitting the error, making an effort to repair the consequences of the error, and then learning what you can from the experience in order to prevent the error in the future.

Exercise 8.11

Return to the Chocolate Bar Study that was described in Example 5.2 in Chapter 5. Put yourself in the place of the statistician for that study. One of your responsibilities is to receive and review the data from the contract laboratory that analyzes the blood samples. In one of the data sets, you are pretty sure that there has been a mixup in the coding of the tubes or some other problem with the assays. This is because values that come from masked replicate tubes are widely divergent from each other. Your first impulse is to send an email note to the entire project team (including the contract laboratory) to announce this problem. Why might this not be such a good idea? What could you do instead?

Exercise 8.12

Return to Dialog 5 in Chapter 5. Suppose that the study had already been conducted and the client has asked you to help at the analysis stage. Develop a script for the way that you would express concern about the client's method of assigning animals to groups. Be sure to balance the "bad news" with some positive comments. Offer options about how to resolve the problems. Feel free to improvise about details of the project.

Exercise 8.13

Return to Dialog 7 in Chapter 5. Suppose that the survey was completed before the client came to you for advice about analysis. As in the previous exercise, develop a script for the way you would express concern about the way the survey was conducted.

Exercise 8.14

Put yourself in the place of the statistician from the video story line. Suppose that after you submitted the final report to Mr. Johnson (the one that will be made public), you discovered that the database had a coding error in it. The computer program mistakenly coded all respondents from India and China as "domestic" rather than "international." As a result, the comparison of the level of satisfaction with UMS services between domestic and international graduate students was wrong and had been misinterpreted in the report. Describe what you would do after discovering this error.

8.10 A Non-Negotiable Position

Suppose your client asks you to do something that you are simply not willing to do. Or, suppose you ask your client to do something that she is not willing to do. You and your client may have tried a number of means to reach an agreement, but to no avail. When this occurs, the two of you are at an impasse. You have reached a non-negotiable position.

You may be in this situation when a client's request has challenged your view of what constitutes good statistical practices. For example, Scott Lasser, the statistician in Example 6.2, was asked to analyze data from a clinical trial repeatedly, without making any adjustments to the error rates of the hypothesis tests. Scott felt that this was not an acceptable statistical practice. He went to the statistics literature, at which point he probably discovered a variety of viewpoints of the seriousness of the problem and what could be done to accommodate it. Statisticians will differ in their interpretation of what they believe to be good statistical practices and a debate on this topic can be very interesting. In fact, participating in such a debate will help you to define your own position on typically controversial issues such as error rates for multiple hypothesis tests, combining information across studies, or defining the scope of inference of a study, to name only a few. You will probably also discover a great deal of overlap between issues of statistical methodology and ethical issues in these controversies.

People in technical professions such as statistics may look to their professional societies for guidance about ethical practices. Both the American Statistical Association and the International Statistical Institute have committees established to develop these guidelines. You can find a wealth of information on the Internet about principles of professional practice in different disciplines. Numerous sites have case studies posted in order to stimulate discussion about ethical issues. A good place to start to find information like this is with the ASA web site (www.amstat.org). Your place of employment may also have written guidelines about professional behavior and practice. You may even wish to develop your own code of professional behavior. W.E. Deming did this for his private consulting practice and then recommended it to the statistics profession (Deming, 1972). In Exercise 8.16, you will have a chance to review and comment on this code.

What should you do when you are asked to do something that is not acceptable to you? You may feel that the best option is to leave the project. In this case, you can use the guidelines from the previous section to help you to deliver this "bad news" in a clear and respectful way. These guidelines will also help you depart in a spirit of good will, so that you have the option (if you wish) of working with this client in the future.

As an illustration, let us return to the story of Scott Lasser in Example 6.2. It was Dr. Sethuramen, a scientist in the field of nutrition, who had asked Scott to analyze the data periodically while the subjects were still accruing into the study. She wanted to be able to stop the study early, as soon as the treatment effects became significant. In the first version of this story, Scott persuaded Dr. Sethuramen to adopt an accepted method for taking these interim looks at the data. In the second version of this story, let us suppose instead that Dr. Sethuramen refused to consider any of the methods for interim analysis that Scott proposed. She continued to ask for periodic unmasked analyses of the data with unadjusted error rates for the hypothesis tests. Let us suppose further that this was unacceptable to Scott. He did not want to continue being associated with a project that made use of this practice. For this reason he decided to withdraw from the project.

This might be a very costly decision for Scott. If he has no real say about which projects are assigned to him, he might have to leave his job. There are many factors that weigh in the balance when it comes to deciding whether or not a position is non-negotiable. However, for purposes of illustration we will assume that Scott has already pursued all other avenues of persuasion and influence. He has offered to give an educational talk so that the scientists at work would better understand this issue. He has tried appealing to people in positions of greater authority in his workplace. He has tried to develop an improved policy on interim analysis that will be in place for future projects. We will assume that he was not satisfied with the outcomes of any of these efforts. We will also assume that he does not want to continue working on this project under these circumstances. Scott then truly has a non-negotiable position.

In this version of the story, Scott has decided to withdraw from the project in a way that will preserve good will. After all, he might need a recommendation from his clients for his next job. When Dr. Sethuramen realizes that he is leaving because of this issue, she might even change her position on interim analysis. With this in mind, Scott has asked to meet in person with Dr. Sethuramen for a private conversation. Exhibit 8.5 contains an account of what he said to her. Scott's language was clear, assertive and non-accusatory. He focused his criticism on the plan, and not on Dr. Sethuramen. He did his best to make this a collegial situation, in which a difference of opinion has led to a cordial decision not to continue working together.

In the second version of Example 6.2, Scott did not appear to have any concerns about Dr. Sethuramen's intent when she asked him to analyze data repeatedly without adjusting for the error rate. Let us move to a third version of this story that has more ethical problems. Suppose in this third version, Scott has discovered that Dr. Sethuramen altered some of the data before giving it to him for analysis. She unmasked the treatment assignments and discovered that some subjects had blood levels that were too low to support the theory that she was trying to prove. She decided to disqualify some of these subjects from the trial. For others, she reasoned that the blood levels must have contained errors. She replaced these levels with values that were more consistent with her theory. In this version of the story, Scott concluded that Dr. Sethuramen had committed fraud.

Exhibit 8.5 ***What Scott said when he withdrew from Dr. Sethuramen's project in the second version of the story***

I appreciate the time and attention you've given this issue of interim analysis in the past few weeks.

...

My understanding from our conversations is that you would still like to continue with the original plan of analyzing the data as the subjects accrue with no adjustment to the statistical tests.

...

This plan does not work for me. It is not consistent with my view of a good statistical practice. I can not implement your preference and still be comfortable with my own development as a knowledgeable statistician. This is why I am withdrawing from this project. I will be glad to provide you with a memo that includes my resignation and outlines the statistical issues that we have already discussed. Perhaps we can work together on some future project where we can agree on some of these basic issues.

What should Scott do? There are several options. He could accept Dr. Sethuramen's behavior and continue with the project. He could leave the project. He could confront Dr. Sethuramen about the fraud. He could report the fraud to someone else. This is a difficult situation indeed. There are professional risks to Scott in this third version of the story. If he continues working on the project, and someone else discovers the fraud, then he might be implicated along with Dr. Sethuramen. If he confronts Dr. Sethuramen about the fraud, Dr. Sethuramen might retaliate. If he reports the fraud to someone else, he may also experience some retaliation. In this situation, Scott would be wise to take measures to protect his professional reputation. He can do so by documenting his actions and making sure that others are present and aware of his interactions with Dr. Sethuramen. He would also benefit from seeking confidential advice, even legal advice, about his options.

Exercise 8.15

Suppose that in the third version of Scott's story, he decided to continue working with Dr. Sethuramen. Scott rationalized to himself that his responsibility extended only to analyzing the data as it was given to him, not in questioning its origins. What do you think of this rationalization? What guidance can you find from ASA's *Ethical Guidelines for Statistical Practice* (available from the ASA web site) on this point of view?

Exercise 8.16

If you were in Scott's position in the third version of the story, would you report Dr. Sethuramen's actions to somebody else? Why or why not? Suppose you had decided to report her actions. How would you go about doing this?

Exercise 8.17

W.E. Deming's Code of Professional Conduct has thirty-seven points organized in seven sections (Deming, 1972). Point #35 is quoted below:

35. Each of us, client and I, has a unilateral right to break off the engagement at any time, with or without explanation. I should feel obligated to break off an engagement if the performance of the investigators or the processing does not in my judgment meet standards that are acceptable for my participation.

If you were developing your own code of professional conduct, would you include this point? Would you keep in the phrase "with or without explanation"? Explain why or why not.

8.11 Unprofessional Behavior

Let us return to segment 8 of the video. In this segment, there is no doubt that Mr. Johnson is angry. Some examples of what he said to Dr. Derr are in Exhibit 8.6. What do you think about Mr. Johnson's statements? He actually expressed his anger in a reasonable sort of way. He used "I" language. He directed his anger at the product (the notebook), not personally at Dr. Derr. We can say that Mr. Johnson stayed within the boundaries of professional behavior even while expressing his anger.

Exhibit 8.6 Mr. Johnson's angry statements

> "This isn't what I asked for."
>
> "I can not use this."
>
> "I do not have time for this."
>
> "We have spent a lot of money for something that I can't use."

Behaving professionally means treating the other person with respect. Respect starts with basic politeness, and extends to the way you communicate and act towards others. The communication skills you have learned in this book are based on respect for the client. Any time you use profanity or slang, or make derogatory generalizations about people, you can offend someone. It is best to be very careful about this type of language or avoid it entirely. Physical contact can also be misinterpreted and give offense. As you have learned in earlier chapters, problems can arise when you and your client have different norms about verbal and non-verbal communication. You may also differ on other points, such as whether or not it is acceptable to be late for meetings. However, if you and your client have good intentions and don't wish to offend each other, you should be able to overcome minor misunderstandings.

What happens when a client oversteps these boundaries? Suppose Mr. Johnson had made the statements shown in Exhibit 8.7. These statements are more likely to give offense than Mr. Johnson's actual remarks. They are accusatory and are directed at Dr. Derr personally. The third remark is a derogatory generalization about women. It would probably be more difficult for Mr. Johnson and Dr. Derr to regain their rapport after these remarks than it was in the actual storyline.

Exhibit 8.7 ***What Mr. Johnson might have said to Dr. Derr***

> "You really messed this up."
>
> "You have wasted all of our money and given us something we can't use."
>
> "It is just typical for a woman not to understand the most important goal of the project."

When a client behaves in a way that you interpret as disrespectful, you have several options. You can ignore the incident. You can let the client know about your reaction, and ask for a change. You can withdraw from working with this client. You can report the incident to someone else. What you decide to do will depend at least in part on how seriously you regard the incident.

Let us return to Example 3.1 from Chapter 3. In this example, Ann Seder was attending a meeting of a project team. At this meeting, Dr. Klerk had grabbed her arm, interrupted her, and then expressed his disagreement with her comments. He did this several times. Ann's first response was to edge her chair away from Dr. Klerk so that he wouldn't be able to reach her. When this did not work, she confronted Dr. Klerk by saying, "I'm not comfortable with the way you are grabbing my arm." This produced an immediate apology from him and an explanation (arm-grabbing was apparently acceptable in his culture). He did not repeat the behavior, and the discussion returned to the scientific issues on the agenda.

If you feel that letting the client know about your reaction is the best option, then the guidelines from the two previous sections can help. Assertive communication about someone's behavior generally follows this sequence: First, describe the behavior. Second, discuss your reactions to this behavior. This can include your views on the consequences of this behavior. Third, make a specific request for a change in behavior. Finally, ask for the client's input about your request.

As an illustration, suppose that Ann from Example 3.1 arranged to meet with Dr. Klerk to discuss his style of interrupting her. The way that her part of the conversation might go is shown in Exhibit 8.8. We don't know whether Dr. Klerk will accept Ann's proposal, but this conversation will help Ann to understand his intentions more clearly. If both Ann and Dr. Klerk would like to work cooperatively with each other, they can probably come up with a way of resolving this difficulty. There are several ways to accomplish this, not just the one way that Ann suggested. For example, Dr. Klerk may explain that this is his style of speaking, and he is not offended when others interrupt him. He may promise to try not to interrupt Ann, and encourage her to interrupt and stop him if he does, with a phrase like "I'm sorry, I wasn't finished yet."

The interaction between Ann and Dr. Klerk took place in the context of a team project. In this context, the team leader can help by establishing the ground rules for discussion. This can be done in a neutral way that does not focus on any one person's style of communicating. The team leader can even designate someone to monitor the discussion and remind people about the ground rules when necessary. This role can rotate to different team members in different meetings. This will help every team member develop an awareness of the dynamics of a discussion and a sensitivity to the needs of each team member.

Exhibit 8.8 ***Example of Assertive Communication***

Ann's comments to Dr. Ulbrecht	Type of assertive communication
Dr. Klerk, I couldn't help but notice that every time I started to speak about the study design, you interrupted me. ...	*Description of client's behavior*
Whenever this happened, I felt frustrated. I felt that I couldn't be clear to the team about my ideas for the design. ...	*Consultant's reactions to client's behavior*
I realize that we have some basic disagreements about this study design, but we won't be able to resolve them very well until everyone on the team hears all of our suggestions. ...	*Consultant's views on consequences of client's behavior*
I would really prefer for each person to be able to finish speaking before the next person starts in.	*Request for change in client's behavior.*
What do you think about this idea? Shall we propose it to the team?	*Request for client's reaction.*

A well-intentioned client and consultant can, given the proper skills, clear up difficulties caused by misunderstandings, different conventions and styles. What happens if you discover that your client does not have these good intentions towards you? Let us look at a second version of the interaction between Ann and Dr. Klerk. In this version, it becomes clear to Ann that Dr. Klerk does not value her input. He feels that she should only be on the team to implement his suggestions, not to contribute her own ideas. He views himself as having much higher status than Ann, and he also holds some negative views about the capacity of people with Ann's demographic characteristics. This is a more difficult situation. Ann may consider leaving the project at this point. She is unlikely to change Dr. Klerk's views of his status or his prejudices about her.

What options does Ann have in this second version? Since this is a team situation, she may be able to seek support from the team leader and any other members who value her participation in a more collegial way. The team leader may recognize that it is his responsibility to make sure that all members are treated with respect. He may also be aware that the productivity of the team will be enhanced when all members feel respected and valued. The team leader may be in a better position than Ann to elicit cooperation from Dr. Klerk. With support from the team leader and other members of the team, Ann may be able to find a way to be an effective consultant and minimize the impact of Dr. Klerk's attitudes.

Now let us look at a third version of this story. In this version, Dr. Klerk is the team leader. Ann has learned that he has made derogatory comments about her ethnic origins. Without her permission, he has telephoned other clients she has worked with and tried to elicit negative comments about her. He has called a team meeting, ordered her not to attend, and asked the other team members to criticize her performance. At other team meetings that she has attended, he has continued to interrupt her. He has opposed every idea that she put forward. In this situation, Ann is at risk professionally. Dr. Klerk is in a position of power and appears to be trying to gather evidence against her. She would be well advised to seek confidential advice, even legal advice, about how to proceed. Dr. Klerk's behavior is outside the boundaries of professional behavior and may even be classified as harassment. In the U.S. and many other countries Ann has legal protection against harassment. In many workplaces there are formal steps that an employee can take if she feels that she is being treated unfairly or is being harassed. Ann may be advised to document Dr. Klerk's communications and behaviors, and to make sure that others have observed them also. She would be wise to make sure that others are present any time he speaks with her. It seems unlikely that this situation will be resolved with Ann and Dr. Klerk finding a way to work cooperatively with each other. Instead, Ann should focus on protecting her professional career.

Let's hope that you are never in this situation, but if it happens, remember that you have the right to be treated with respect. It would be very difficult to interact effectively about the technical issues of a project in such a negative environment as the one described in the third version of this story. The best you can do is approach each client and each project team with the respect that you would also like to receive. Then remain attuned to the situation and be ready to respond if a problem arises. A good strategy is to treat an emerging problem first as a minor misunderstanding between well-intentioned people. If your attempts to resolve this misunderstanding reveal something more serious, then some of the methods covered in this chapter should help you address them. These methods are most likely to help lead to a resolution between people who really do want to work together. If someone does not have good intentions towards you, a good plan is to balance your efforts between protecting yourself professionally, holding the person accountable for his behavior and promoting a resolution.

Exercise 8.18

Put yourself in the place of the statistician in segment 8 of the video story line. Suppose that Mr. Johnson made the remarks to you shown in Exhibit 8.7. Write out your comments to him that you would use to restore the spirit of collegiality between the two of you.

Exercise 8.19

Develop a first draft of a description of good professional behavior for a statistical consultant and for a client. Show your first draft to a client and incorporate her suggestions into your revision. Comment on the client's reactions to your first draft.

Exercise 8.20

Develop a fourth version of the story of Ann and Dr. Klerk. In this version, make Ann the team leader. Suggest what Ann can do to address his actions, as they were described in the second version of the story and as they were described in the third version of the story.

8.12 Suggestions for Group Discussion

1. Organize class discussion around some of the exercises so that people can appreciate the different points of view in the class. You can also break the class into smaller groups to work on some of the exercises. Have a spokesperson report on the small group discussion.

2. Experienced statistical consultants and clients can contribute a great deal to the discussion of certain exercises, such as 8.4-8.6, 8.10, 8.14 – 8.17, 8.19, 8.20. You can assign one to each small discussion group, or else have them address the entire class.

3. Invite experienced statistical consultants to contribute descriptions of difficult situations. These can be used in small discussion groups. You can ask the group to characterize each situation (see Exhibit 8.2) and brainstorm some solutions.

4. Organize a brain storming session around Exercises 8.4 or 8.5, to give your class experience in this problem solving technique.

5. If you have videotapes of consulting sessions, examine these for examples of consulting breakdowns. You can use these tapes to coach your students in how to recognize, avoid, and resolve breakdowns. See Zahn (1988) for suggestions about coaching with videotapes. You can also invite clients to view these videotapes with the class and provide their own interpretations. It would be interesting to hear about the negative attributions that clients may make about statisticians during the process of a breakdown.

6. You can expand the discussion of ethical conduct and good statistical practices. These are important topics in statistical consulting and are not often covered in traditional statistics courses. There is a great deal of information on the Internet. The ASA web site is a good starting place. You can organize a discussion around case studies referenced on the Internet, stories in the news, or the experiences from your class. Invite experienced statisticians and clients to participate in the discussion.

7. You can develop Exercise 8.19 into a class exercise, with everyone contributing to the first draft. Then arrange to have one or more clients critique this draft. The exchange of views about good professional behavior should make for good discussion.

8. Many difficult situations in statistical consulting arise due to a lack of team skills. In traditional statistics courses, students don't get much opportunity to learn how to work effectively in a team environment. You may be able to arrange for your students to work in interdisciplinary project teams. If so, be sure to discuss team skills and evaluate students accordingly.

8.13 Resources

Cohen, A.R. and Bradford, D.L. (1991), *Influence without Authority,* NY: John Wiley & Sons.

Deming, W.E. (1972), "Code of professional conduct," *International Statistical Reviews,* 40, 215-219.

Ross, N. P. (1995), "What the government needs," *The American Statistician,* 49, 7-9.

Rusk, T. (1993), *The Power of Ethical Persuasion: From Conflict to Partnership at Work and in Private Life,* NY: Penguin Books.

Yeatts, D.E. and Hyten, C. (1998), *High-Performing Self-Managed Work Teams: A Comparison of Theory to Practice,* Thousand Oaks, CA: SAGE Publications.

Zahn, D.A. (1988), "Quality breakdowns: An opportunity in disguise," *1988 ASQC Quality Congress Transactions*, 56-62.

Zahn, D.A. and Boroto, D.R. (1983), "Being responsible for the success of consulting sessions," *American Statistical Association 1983 Proceedings on Statistical Education*, 107-108.

9

CONCLUSION

Do you feel ready to apply the communication skills that you have learned in this book to your work with clients? There is a lot to keep in mind once you get involved in a meeting. It is fine to learn about communication skills one at a time, but once you start interacting with a client, you will need to integrate them into a coherent whole. There is no better way to gain competency in communication skills than by consulting with actual clients. Even if you have consulted before, you may be attempting to modify your style and incorporate some of the suggestions that you have read about in this book. Here are some suggestions to help you on your way to integrating new skills in your work with clients:

Assess your strengths and weaknesses. It may help to get a sense of where you are right now with these skills. This can serve as a baseline from which to track your progress. At the end of this chapter is a set of assessment forms, one for each of Chapters 3 through 8. Each form identifies the communication skills and related strategies that were covered in that chapter. You can use these forms to guide your self-assessment. You can repeat this assessment later on as a way of identifying skills that you are satisfied with and skills that still need improvement.

Find someone to help you. This could be an instructor, a colleague at work, or someone else who is willing to give you feedback and suggestions about your consulting skills. Your "coach" does not need to be a statistician. Someone who is attuned to the dynamics of meetings and communication in general can give you good advice. You may be able to provide your coach with data by recording a session or two with a client. You can also ask your coach to evaluate your skills by filling out the assessment forms at the end of this chapter.

Start small. It can be overwhelming to make a lot of changes at once to the way that you communicate with others. You might not even be very successful with a vast and sweeping change. This is partly why we looked at sets of communication skills one at a time and practiced them in separate chapters of this book. Once you begin working with clients, consider focusing on one skill at a time. You can start by selecting a skill that you don't think will be too difficult to improve. After you have developed some mastery with this skill, you can build others into your improved style one by one. Consider approaching them in order from "easy" to "difficult" from your perspective. For example, I first worked on asking better questions. I noticed a very rapid improvement in the quality of information I received from a client once I learned to ask more open questions and reflected my understanding back to the client with concrete paraphrases. What was more difficult for me was to converse effectively with someone who had a pronounced circular style of ordering information. This is because I have a very pronounced linear "start to finish" style. However, with time and practice, I became better at doing this. The next most difficult skill for me was presenting technical information to people with very different learning styles than I have. I have a strong preference for the visual modality for learning, and I am comfortable speaking in generalities and abstractions. Sometimes I forget to be more specific and concrete. Again, with increased awareness of my own style and the diversity of other styles, I have learned to pick up more cues about my client's preferences and adapt my presentation accordingly. Finally, I have the greatest difficulty in recognizing, defusing, and resolving certain types of difficult situations – so you can say I am still working on my communication skills!

Reflect on the experience. At key points in the consultation, take time to reflect on how things are going. You might want to do this after each meeting, after a project milestone has been reached, or at the end of a project. Ask yourself whether you feel you are being effective. Recall that effectiveness, at least in this book, is defined by two outcomes: (1) you and your client are both satisfied with the consultation; and (2) you have persuaded your client to make use of good statistical practices. If you have different criteria for effectiveness, then use those. What is important is that you think about these outcomes, identify any roadblocks to effectiveness that you have encountered, develop a strategy to address the roadblocks, and list what you have learned that you might apply in the next meeting or the next project.

Seek feedback from your client. Your client is in a good position to let you know how things are going. He may have some suggestions for how to improve the experience.

Don't demand perfection from yourself. Very few people will be excellent at all of the communication skills that were covered in this book. It is not necessary to be excellent in all skill areas in order to be an effective statistical consultant. Your intent to be effective will probably be very evident to your client, and this can go a long way towards promoting a successful consulting experience.

Exhibit 9.1 ***Assess your communication skills from Chapter 3 "First Contact, Firs[t] Impressions"***

	Communication Skills and Related Strategies	EXCEL-LENT	GOOD	NEEDS SOME WORK	NEEDS A LOT OF WORK
3.1	Providing a welcoming setting for a meeting	☐	☐	☐	☐
3.2	Starting with a welcoming greeting	☐	☐	☐	☐
3.3	Accommodating the client's style of greeting	☐	☐	☐	☐
3.4	Communicating your interest through your eye contact	☐	☐	☐	☐
3.5	Interpreting your client's use of eye contact	☐	☐	☐	☐
3.6	Adjusting your inter-personal distance to accommodate your and your client's preferences	☐	☐	☐	☐
3.7	Communicating your interest through your body language	☐	☐	☐	☐
3.8	Interpreting the non-verbal communication in a group setting	☐	☐	☐	☐
3.9	Building rapport over the telephone	☐	☐	☐	☐

Comments about the skill assessment from Chapter 3:

Exhibit 9.2 ***Assess your communication skills from Chapter 4 "Meetings"***

	Communication Skills and Related Strategies	Excel-lent	Good	Needs some work	Needs a lot of work
	If you are meeting one-on-one with a client:				
4.1	Using cues to identify the client's preference for phasing in "small talk"	☐	☐	☐	☐
4.2	Using cues to identify the client's preferences for sequencing information	☐	☐	☐	☐
4.3	Setting an agenda with your client	☐	☐	☐	☐
4.4	Accommodating your and your client's preferences for sequencing	☐	☐	☐	☐
4.5	Bringing your meeting to closure	☐	☐	☐	☐
4.6	Summarizing your discussion	☐	☐	☐	☐
	If you are participating in a team meeting:				
4.7	Participating in rapport-building activities	☐	☐	☐	☐
4.8	Contributing to the discussion in an in-person meeting	☐	☐	☐	☐
4.9	Contributing to a teleconference discussion	☐	☐	☐	☐
4.10	Contributing to a video conference discussion	☐	☐	☐	☐
	If you are the discussion leader of a team meeting:				
4.11	Providing for rapport-building activities	☐	☐	☐	☐
4.12	Developing an agenda	☐	☐	☐	☐
4.13	Allowing for the preferences of individuals while following the agenda	☐	☐	☐	☐
4.14	Leading a teleconference	☐	☐	☐	☐
4.15	Leading a video conference	☐	☐	☐	☐

Communication Skills and Related Strategies	EXCEL-LENT	GOOD	NEEDS SOME WORK	NEEDS A LOT OF WORK
4.16 Bringing a discussion to closure	☐	☐	☐	☐
4.17 Summarizing the discussion	☐	☐	☐	☐

Comments about the skill assessment from Chapter 4:

Exhibit 9.3 ***Assess your communication skills from Chapter 5 "Asking Good Questions"***

	Communication Skills and Related Strategies	EXCEL-LENT	GOOD	NEEDS SOME WORK	NEEDS A LOT OF WORK
5.1	Identifying whether the client's study is a designed experiment, a survey, an observational study, some mix of the three, or some other type of study	☐	☐	☐	☐
5.2	Identifying the stage of the client's study	☐	☐	☐	☐
5.3	Listing what you should find out about the client's study	☐	☐	☐	☐
5.4	Finding ways to learn more about your client's field of application	☐	☐	☐	☐
5.5	Using open probes to get general information	☐	☐	☐	☐
5.6	Using concrete paraphrases to clarify your understanding	☐	☐	☐	☐
5.7	Using closed probes to get specific information	☐	☐	☐	☐
5.8	Using open probes to signal the transition between topics	☐	☐	☐	☐
5.9	Using cues to identify how your client prefers to organize general and specific information	☐	☐	☐	☐
5.10	Adapting the discussion to suit both your and your client's preferences for specificity	☐	☐	☐	☐

Comments about the skill assessment from Chapter 5:

Exhibit 9.4 ***Assess your communication skills from Chapter 6 "Negotiating a Satisfactory Exchange"***

	Communication Skills and Related Strategies	Excel-lent	Good	Needs some work	Needs a lot of work
6.1	Including negotiation early in your work with the client	☐	☐	☐	☐
6.2	Identifying the issues of a negotiation	☐	☐	☐	☐
6.3	Characterizing the positions held by you and your client	☐	☐	☐	☐
6.4	Identifying the tangible items of an exchange	☐	☐	☐	☐
6.5	Identifying the intangible items of an exchange	☐	☐	☐	☐
6.6	Assessing the relative value that the client places on each item of an exchange	☐	☐	☐	☐
6.7	Identifying the relative value that you place on each item of an exchange	☐	☐	☐	☐
6.8	Assessing the extent to which tangible and intangible items can be substituted for each other	☐	☐	☐	☐
6.9	Using cues to identify whether a client has a high context or low context style of negotiation	☐	☐	☐	☐
6.10	Using cues to identify your client's preferences for objectivity (directness vs. indirectness)	☐	☐	☐	☐
6.11	Adapting the discussion to suit both your and your client's preferred styles	☐	☐	☐	☐
6.12	Characterizing the outcomes of the negotiation (such as: win-win, win-lose, lose-lose)	☐	☐	☐	☐

Comments about the skill assessment from Chapter 6:

Exhibit 9.5 Assess your communication skills from Chapter 7 "Communicating About Statistics"

	Communication Skills and Related Strategies	EXCEL-LENT	GOOD	NEEDS SOME WORK	NEEDS A LOT OF WORK
	For a one-on-one discussion:				
7.1	Using cues to identify your client's preferred style(s) of learning	☐	☐	☐	☐
7.2	Adapting your discussion to your client's preferred style(s) of learning	☐	☐	☐	☐
7.3	Clearly defining statistical terms and concepts	☐	☐	☐	☐
7.4	Assessing how much your client understands of the statistical content of your discussion	☐	☐	☐	☐
	For a presentation:				
7.5	Understanding the needs and expectations of your audience	☐	☐	☐	☐
7.6	Addressing your audience's objectives	☐	☐	☐	☐
7.7	Staying within your time limit	☐	☐	☐	☐
7.8	Making use of excellent visuals	☐	☐	☐	☐
7.9	Seeking input from your audience as part of your preparation	☐	☐	☐	☐
7.10	Providing a handout that reinforces your talk	☐	☐	☐	☐
	For a document:				
7.11	Identifying the audience	☐	☐	☐	☐
7.12	Creating a succinct document for a busy reader	☐	☐	☐	☐
7.13	Making it easy for the reader to navigate through the document	☐	☐	☐	☐

Communication Skills and Related Strategies	EXCEL-LENT	GOOD	NEEDS SOME WORK	NEEDS A LOT OF WORK
7.14 Making good use of tables and graphs	☐	☐	☐	☐
7.15 Seeking input from your clients while preparing the document	☐	☐	☐	☐

Comments about the skill assessment from Chapter 7:

Exhibit 9.6 ***Assess your communication skills from Chapter 8 "Difficult Situations"***

	Communication Skills and Related Strategies	Excel-lent	Good	Needs some work	Needs a lot of work
8.1	Using your negative attributions about the client only as a cue that a breakdown might be taking place	☐	☐	☐	☐
8.2	Characterizing the difficult situation	☐	☐	☐	☐
8.3	Identifying long-term actions that can alleviate a difficult situation	☐	☐	☐	☐
8.4	Using cues to identify whether a team conflict is cooperative or competitive	☐	☐	☐	☐
8.5	Taking steps to be regarded as a valuable team member	☐	☐	☐	☐
8.6	Promoting a positive team environment	☐	☐	☐	☐
8.7	Delivering "bad news" to the most appropriate person(s)	☐	☐	☐	☐
8.8	Choosing an appropriate setting for a discussion of "bad news"	☐	☐	☐	☐
8.9	Using appropriate language in discussing the problem	☐	☐	☐	☐
8.10	Expressing your concern in terms of the goals of the project	☐	☐	☐	☐
8.11	Balancing the negative with the positive	☐	☐	☐	☐
8.12	Offering options	☐	☐	☐	☐
8.13	Taking responsibility for your errors	☐	☐	☐	☐
8.14	Identifying positions that are non-negotiable	☐	☐	☐	☐
8.15	Using assertive communication when requesting a change in the client's behavior	☐	☐	☐	☐

Comments about the skill assessment from Chapter 8:

Index